MW01199679

An Introduction to Linear Programming and the Theory of Games

Abraham M. Glicksman

DOVER PUBLICATIONS, INC.
Mineola, New York

Copyright

Published in Canada by General Publishing Company, Ltd., 30 Lesmill Road, Don Mills, Toronto, Ontario.

Bibliographical Note

This Dover edition, first published in 2001, is an unabridged reprint of the work, originally published by John Wiley & Sons, Inc., New York, in 1963.

Library of Congress Cataloging-in-Publication Data

Glicksman, Abraham M.
An introduction to linear programming and the theory of games / Abraham M. Glicksman.
p. cm.
Originally published: New York : J. Wiley, 1963.
Includes bibliographical references and index.
ISBN 0-486-41710-7 (pbk.)
1. Linear programming. 2. Game theory. I. Title.

QA269 .G55 2001
519.7'2—dc21

00-065609

Manufactured in the United States of America
Dover Publications, Inc., 31 East 2nd Street, Mineola, N.Y. 11501

To Harold W. Kuhn and Albert W. Tucker

Preface

This monograph introduces the reader to linear programming and the theory of games in a strictly elementary yet mathematically sound manner. It goes beyond the usual elementary graphical treatment of linear inequalities. It also exposes, discusses, and derives some of the deeper algebraic ideas and theorems. These include *the Fundamental Extreme Point Theorem* for convex polygons, Dantzig's *Simplex Method, the Fundamental Duality Theorem,* and von Neumann's *Minimax Theorem.*

These formidable-sounding ideas are introduced in a concrete and leisurely manner, using only techniques already well known to a good high school student or college undergraduate. Nevertheless, the language is precise, the arguments quite general, and the reasoning completely sound.

I believe that this is by far the simplest nontrivial exposition of linear programming and the theory of games which has yet appeared. It is eminently suitable for high school "honors" programs, for college students, for teachers, and for interested laymen.

In writing this book, I drew inspiration from so many sources that it is now impossible to acknowledge them all. At best, I can but mention a few and hope that the many others will understand.

I am especially grateful to Harold W. Kuhn and A. W. Tucker of Princeton, whose interest, advice, and encouragement were as important to me as their generous supply of up-to-date material and accurate information. George B. Dantzig of the University of

California was also kind enough to furnish me with a number of valuable reprints.

From my students at the Bronx High School of Science as well as my colleagues, I drew many excellent ideas for the formulation of problems and presentation of material. Particular appreciation is due colleagues Louis Cohen, Charles Hodes, and Harry Ruderman, who read the manuscript. Their valuable comments and criticisms eliminated many errors and ambiguities. Several of the problems used in the text were adapted from materials supplied by Samuel L. Greitzer of Rutgers University and Irving A. Dodes of the Bronx High School of Science.

To these and many other people, I extend my sincerest thanks.

December, 1962 A. M. GLICKSMAN

Contents

PART ONE

Elementary Aspects of Linear Programming

1 Preliminary Remarks

In 1947 a lusty mathematical infant appeared in the United States. At that time, George B. Dantzig and his associates in the U.S. Department of the Air Force recognized a common element in a wide variety of military programming and planning problems. Present in all of them was the purely mathematical task of maximizing or minimizing a *linear form* whose variables were restricted to values satisfying a system of *linear constraints* (i.e., a set of linear equations and/or inequalities). It was quite natural to designate this mathematical problem by the term *linear programming*.

Curiously enough, in 1939, a Russian mathematician, L. V. Kantorovich and his associates had been studying a similar set of problems, but this work was unknown to the Americans. Working independently of the Russians, Dantzig developed a systematic procedure for solving linear programming problems. This procedure, the *Simplex Method*, came to be recognized as the most effective general method for handling linear programming problems. Subsequent research went into painstaking effort to improve and refine the Simplex Method to make it a precise and reliable mathematical tool. As an *iterative* procedure it was readily adapted for use on modern high-speed electronic computers. It became an important technique in theoretical investigations, and its applications expanded into industry, agriculture, transportation, economic theory, and engineering—a remarkable performance record indeed, for such a mathematical infant.

New ramifications of linear programming are continually coming to light. One of the most interesting of these ramifications was the early discovery of the close connection of linear programming with the Theory of Games. Originally proposed about 1921, by the French mathematician Émile Borel, this theory received a strong impetus in 1928 when its key result, the *Minimax Theorem*, was first proved by John von Neumann. With the subsequent development of linear programming, the difficult and sophisticated theory of games underwent considerable simplification. Research by G. B. Dantzig, L. S. Shapley, A. J. Hoffman, Philip Wolfe, David Gale, A. W. Tucker, H. W. Kuhn, S. Vajda, M. Beale, A. Charnes, and others have brought both game theory and linear programming to a high state of perfection.

It is the object of this monograph to introduce the reader to the central ideas involved in linear programming and the theory of games in such a manner as to reveal the essentially elementary character of these theories. The techniques used throughout are those already familiar to a good high school student or college undergraduate. The only additional requisite is the ability to follow a logical argument.

We shall begin by analyzing carefully several (oversimplified) problems. These will serve to introduce the underlying ideas and some of the terminology. The reader will also obtain a new "slant" on a number of classical techniques in algebra and geometry with which he is undoubtedly already familiar. As we move on to examine generalizations which naturally suggest themselves, there will emerge in a fairly elementary, yet rigorous manner, the

Fundamental Extreme Point Theorem, the Simplex Method, the Fundamental Duality Theorem, and its corollary the Minimax Theorem. A number of problems are solved in detail using both extended and condensed tableaux, and further exercises (with answers indicated) are included for the ambitious reader.

2 Graphical Solution of Linear Programming Problems

As our first example of a (highly oversimplified) linear programming problem, let us study the following:

Example I. Dr. Sy Koso Ma Teek, the eminent medical specialist, claims that he can cure colds with his revolutionary 3-layer pills. These come in two sizes: *regular size* containing 2 grains of aspirin, 5 grains of bicarbonate, and 1 grain of codeine; *king size* containing 1 grain of aspirin, 8 grains of bicarbonate, and 6 grains of codeine. Now, Dr. Sy Koso's research has convinced him that it requires at least 12 grains of aspirin, 74 grains of bicarbonate, and 24 grains of codeine to effect his remarkable cure. Determine the *least* number of pills he should prescribe in order to meet these requirements.

Analysis and Solution

Let x = the number of *regular-size* pills prescribed.
y = the number of *king-size* pills prescribed.

The problem imposes *constraints* (restrictions) upon the values of x and y, some of which are explicitly stated and some implied by the nature of the problem. For example, it makes no sense to assign negative values to x or y, so we must have

$$x \geq 0 \quad \text{and} \quad y \geq 0.$$

We observe next that x regular-size pills will provide $2x$ grains of aspirin while y king-size pills provide $1y$ grains of aspirin and these two amounts combined must aggregate *at least* 12 grains of aspirin. Accordingly we obtain the following constraint:

$$2x + 1y \geq 12.$$

Similarly, the aggregate amount of bicarbonate must amount to *at least* 74 grains, and this requirement is expressed as follows:

$$5x + 8y \geq 74.$$

A final constraint is imposed by the specification of *at least* 24 grains of codeine as a requisite for the cure:

$$1x + 6y \geq 24.$$

We shall see that there are many possible values for x and y which will satisfy all these restrictions, but our problem is to select from among these a pair of values that will make the total number of pills, namely $x + y$, *as small as possible.* Let us designate this total number of pills by m. Then we may restate our problem in the following mathematical form, typical of linear programming problems generally:

Determine: $x \geq 0$ and $y \geq 0$

so that: $\begin{cases} 2x + 1y \geq 12 \\ 5x + 8y \geq 74 \\ 1x + 6y \geq 24 \end{cases}$

and so that: $m = x + y$ is a MINIMUM.

We proceed now to examine each of the constraints more carefully. Consider the very first one, $x \geq 0$. Stated more precisely, this means that *either* $x = 0$ *or* $x > 0$. Graphically, the solution set for $x = 0$ consists of all points whose coordinates are of the form $(0, y)$, where y may be any real number. Clearly these points all lie on the y-axis (see Figure 1).

The solution set for the constraint $x = 0$ corresponds to the set of all points on the y-axis.

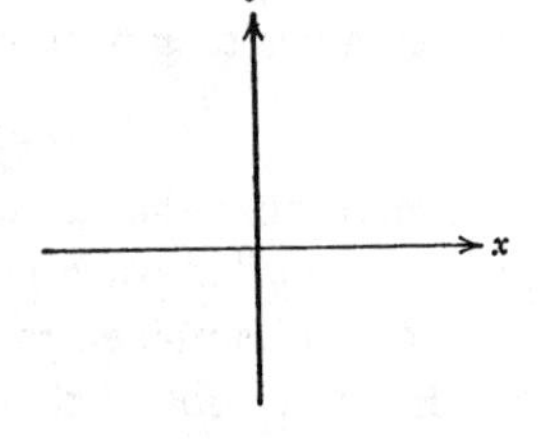

Figure 1

The constraint $x > 0$ corresponds graphically to the set of all points whose coordinates are of the form (p, y), where p can be any *positive* real number. All such points are located to the "right" of the y-axis and make up the "right half" of the xy-plane (see Figure 2). Such a set is called an *open half plane.*

The solution set for the constraint $x > 0$ corresponds to the set of all points to the right of the y-axis.

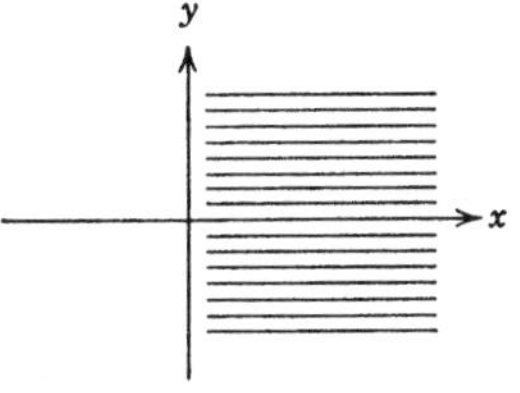

Figure 2

The constraint $x \geq 0$ therefore corresponds graphically to the *union* of these two sets of points, i.e., to all points which are located *either* on *or* to the right of the y-axis (see Figure 3). Such a set is called a *closed half plane.*

The solution set for the constraint $x \geq 0$ corresponds to the set of all points either on or to the right of the y-axis.

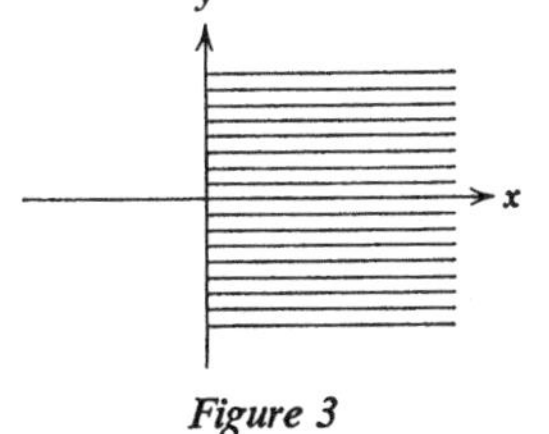

Figure 3

The next constraint appearing in our problem is $y \geq 0$. By a similar argument, its solution set is seen to correspond to the closed half plane which consists of all points either on or above the x-axis (see Figure 4).

The solution set for the constraint $y \geq 0$ corresponds to the set of points on or above the x-axis.

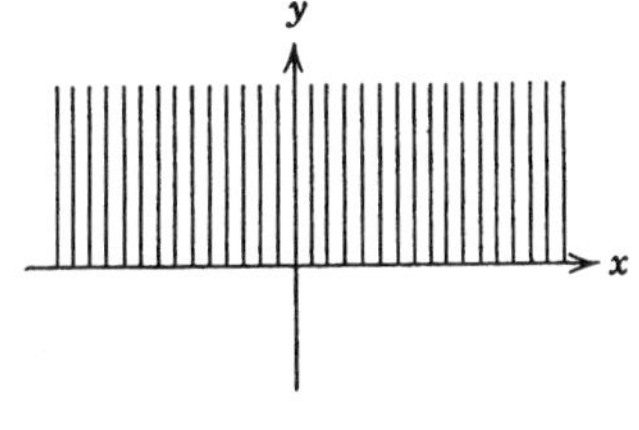

Figure 4

However, the solutions we are seeking must satisfy *both* $x \geq 0$ *and* $y \geq 0$. Consequently, they must correspond graphically to points which are contained in both of the sets depicted in Figures 3 and 4. The collection of points common to both of these sets is called their *intersection.* It is shown in Figure 5.

The intersection of the two closed half planes corresponding to the constraints $x \geq 0$ and $y \geq 0$

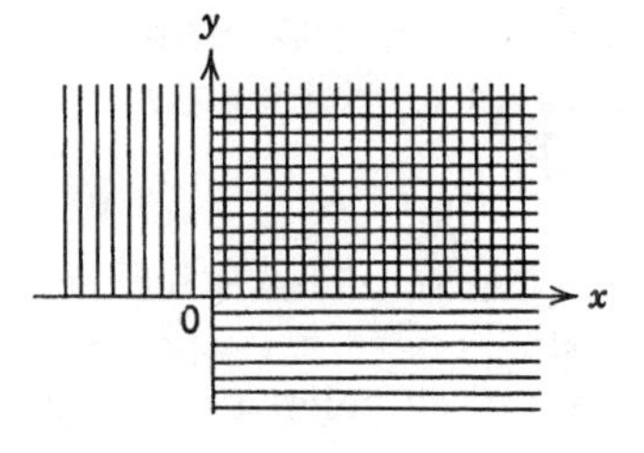

Figure 5

Observe that the "non-negativity requirements" $x \geq 0$ and $y \geq 0$, confine the solutions to points within (or on the "boundary" of) the first quadrant. Our problem is to select from this infinite set, points corresponding to the further constraints

$$2x + 1y \geq 12$$

$$5x + 8y \geq 74$$

$$1x + 6y \geq 24.$$

Consider the first of these inequalities. It stands for

$$2x + y = 12 \quad or \quad 2x + y > 12.$$

The solution set for $2x + y = 12$ corresponds to all points on a straight line l defined by $y = -2x + 12$. This line has a slope of

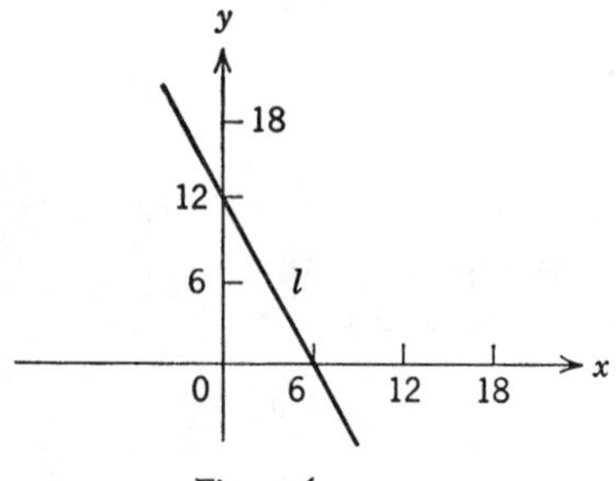

Figure 6

-2 and a y-intercept of 12. It is thus readily graphed as in the adjoining Figure 6. Now, if we let S denote the set of points whose coordinates satisfy $2x + y > 12$, then S will consist of all points which are *above* the line l. A convincing argument to this effect follows: let x be held fixed, let $P = (x, y)$ be a point of S, and let $Q = (x, y')$ be a point of l (both P and Q have the same first coordinate x). Now

$$2x + y > 12, \quad \therefore y > -2x + 12;$$

$$2x + y' = 12, \quad \therefore y' = -2x + 12.$$

$$\therefore y > y', \text{ i.e., } P \text{ lies above } Q!$$

(See Figure 7.)

No matter what point P is chosen in S, there will therefore exist on line l a point Q, having the same abscissa as point P; furthermore, P will always be situated above Q. This is what one means by the assertion, "all points of S are above line l."

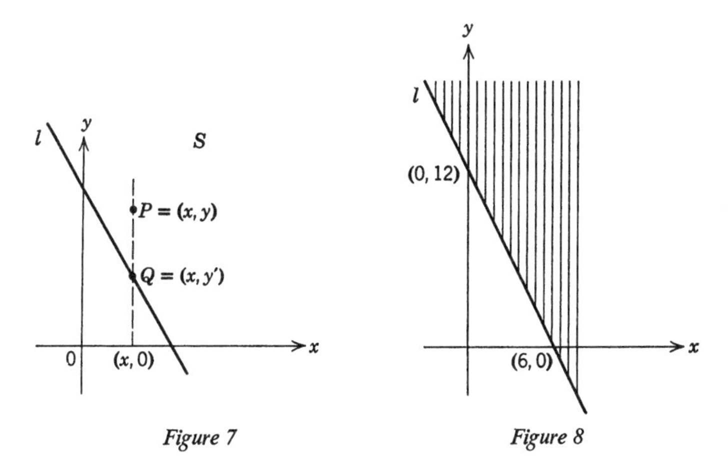

Figure 7 *Figure 8*

It should now be evident that the solution set for the constraint $2x + 1y \geq 12$ corresponds to the set of all points *on or above* the line l, defined by the equation $2x + 1y = 12$ (see Figure 8).

We digress to remark that this last set of points is another example of a *closed half plane*. The line l itself is called the *boundary* of this closed half plane. If the boundary is removed, the remaining points (in this case those which lie entirely above line l) form an *open half plane*.

Let us return now to the remaining constraints of our problem. Each of these constraints defines a closed half plane also (see Figures 9 and 10).

Solution set for the constraint $5x + 8y \geq 74$

Solution set for the constraint $1x + 6y \geq 24$

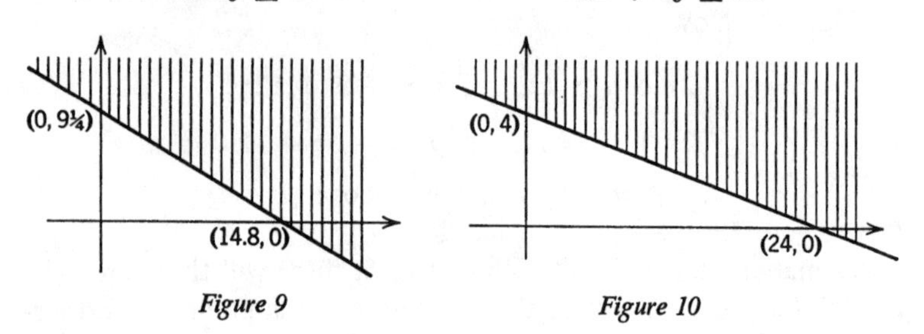

Figure 9

Figure 10

If there are any solutions at all to our problem, these solutions must correspond to points which are contained within *all* of these half planes. Thus each constraint defines a certain half plane, and any solution to the problem (if one exists) defines a point belonging to the *intersection* of all these half planes. This intersection is depicted in Figure 11. (The arrows point from the *boundary* into the *interior* of the intersection. This is less cumbersome than shading the region.)

Intersection of the closed half planes defined by the constraints:

$$x \geq 0, y \geq 0,$$
$$2x + 1y \geq 12,$$
$$5x + 8y \geq 74,$$
$$1x + 6y \geq 24.$$

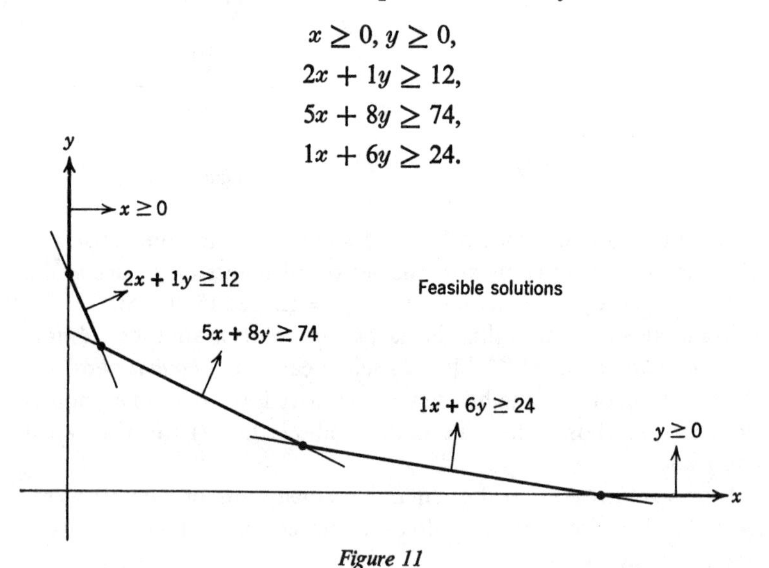

Figure 11

Any member of this intersection represents what is called a *feasible solution.* We shall also call this part of the graph the *feasible region* for our linear programming problem. Our task is to select from among all these feasible solutions one which makes

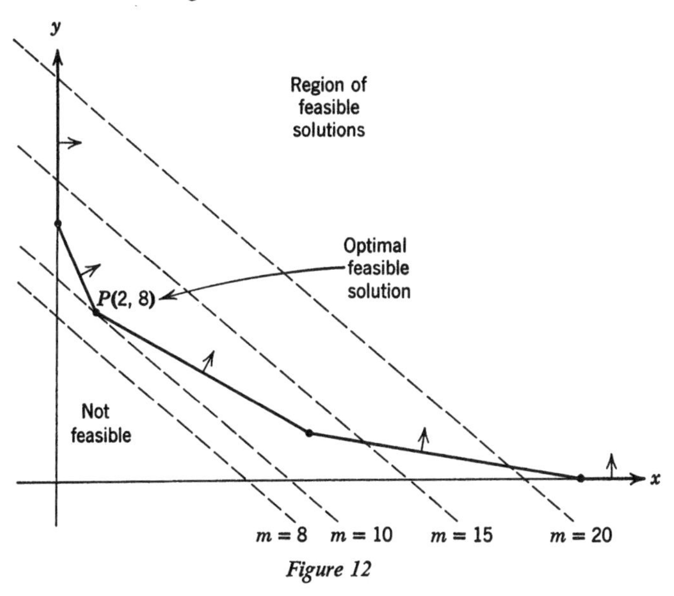

Figure 12

$x + y$ a minimum. In the present case this can be accomplished in a fairly simple, "common sense" way. Let us write

$$x + y = m$$

and assign different values to the symbol m. For example we might write

$$x + y = 20$$
$$x + y = 15$$
$$x + y = \ \ 8$$
etc.

We can then plot these equations on our graph as a *family* of parallel straight lines. (They are parallel because they all have the same slope -1.) Wherever one of these lines crosses the feasible region on our graph, the coordinates x and y form a feasible solution, for which the sum $x + y$ is equal to the value assigned to m, for that particular line (see Figure 12).

We observe next that for some values of m there are no feasible solutions at all. This occurs for all values of m less than 10. When $m = 10$, the line just "grazes" the feasible region at point P, where $x = 2$ and $y = 8$. For all larger values of m (more than 10), the line penetrates the feasible region and yields feasible solutions for which the sum $x + y$ is larger than 10. Clearly, the *optimal* (best) solution to this particular problem occurs at the point P, where $x = 2$, $y = 8$ and therefore, $m = 10$. A prescription of 2 *regular-size* and 8 *king-size* pills represents the *least* number of pills which Dr. Sy Koso Ma Teek can specify as meeting all the requirements. Incidentally, this prescription exactly meets the minimum specifications for aspirin and bicarbonate, but it yields a substantial "overdose" of codeine, beyond the minimum requirement, namely $1(2) + 6(8) =$ 50 grains (which is considerably more than the minimum dose of 24 grains actually needed!). This "waste" is often referred to as *slack*. It is unavoidable in this problem and can only be reduced by increasing the total number of pills.

Certain features of this problem are worth a bit of further discussion, for they are quite general. The region of feasible solutions consisted of all points in the *intersection* of a finite number of half planes. Such a region is often referred to as a *polygonal region*, or *polygonal set*. The optimum solution was found to be situated on the *boundary* of this polygonal region, in fact at a *vertex* (corner) where at least two boundary lines came together. In the next chapter we shall verify that these features are indeed quite general, but before we do this let us examine another specific problem.

Example 2. A confectioner manufactures two kinds of candy bars, ERGIES (packed with energy for the growing kiddies) and NERGIES (the "no-cal" nugget for weight watchers without will power). ERGIES sell at a profit of 40¢ per box, while NERGIES bring in a profit of 50¢ per box. The candy is processed in three main operations, *blending*, *cooking*, and *packaging*. The following table records the average time in minutes, required by each box of candy, for each of the three processing operations:

	Blending	Cooking	Packaging
ERGIES	1 min.	5 min.	3 min.
NERGIES	2 min.	4 min.	1 min.

During each production run, the blending equipment is available for a maximum of 12 machine hours, the cooking equipment for at most 30 machine hours, and the packaging equipment for no more than 15 machine hours. If this machine time can be allocated to the making of either type of candy at all times that it is available, determine how many boxes of each kind the confectioner should make in order to realize the MAXIMUM profit.

Analysis and Solution

Let $x =$ the number of boxes of ERGIES to be made.
$y =$ the number of boxes of NERGIES to be made.

As in the previous problem, negative values of x and y must be excluded, i.e.,

$$x \geq 0 \quad \text{and} \quad y \geq 0.$$

The total blending time, in minutes, required for the processing of x boxes of ERGIES and y boxes of NERGIES is $1x + 2y$ and this may not exceed $(12)(60) = 720$ minutes. Hence

$$1x + 2y \leq 720.$$

The limitation of 30 machine hours, at most, on the cooking equipment is expressed by the constraint

$$5x + 4y \leq 1800.$$

The packaging equipment limitation takes the form

$$3x + 1y \leq 900.$$

Subject to these restrictions the value of the profit must be maximized, namely $40x + 50y$. Let us designate the value of the profit by M. We may restate our problem in the following (typical) form:

Determine: $x \geq 0$ and $y \geq 0$

so that:
$$\begin{cases} 1x + 2y \leq 720 \\ 5x + 4y \leq 1800 \\ 3x + 1y \leq 900 \end{cases}$$

and so that: $M = 40x + 50y$ is a MAXIMUM.

As in the previous problem, the non-negativity constraints $x \geq 0$, $y \geq 0$ confine the feasible solutions to the first quadrant. The constraint $x + 2y \leq 720$ is represented by a closed half plane consisting of all points on, or *below*, the line defined by $x + 2y = 720$. Thus far, our graph appears as in Figure 13. If we intersect

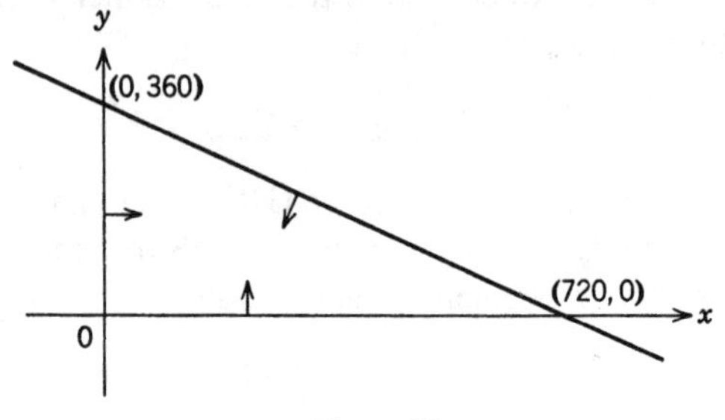

Figure 13

this triangular region with two further half planes, corresponding to the two remaining constraints, we obtain a region of feasible solutions for our problem. This region appears as a *closed polygon* (see Figure 14).

Now, once again our problem is to select from among the infinitely many points of this feasible region an "optimal" point, i.e.,

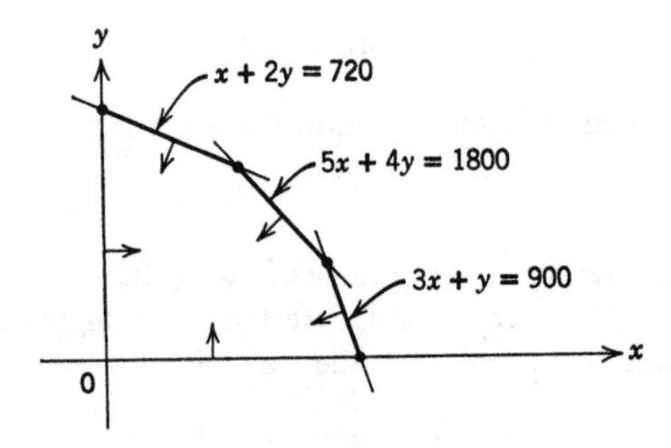

Figure 14

one corresponding to values of x and y which will maximize the value of the profit M. We accomplish this by graphing the profit formula $40x + 50y = M$, for various values of the parameter M. As in the previous problem, some values of M yield straight lines entirely outside of the feasible region, while other values of M correspond to lines which intersect this region. All the lines in this "family" are parallel to each other, as they all have the same slope, namely $-\frac{4}{5}$, One of these lines just "grazes" the region at point P, indicated in Figure 15. The coordinates of P are readily obtained by solving the pair of equations $x + 2y = 720$ and $5x + 4y = 1800$. We find

$$x = 120 \quad \text{and} \quad y = 300,$$

$$M = 40(120) + 50(300) = 19{,}800,$$

indicating that a production program of 120 boxes of ERGIES and 300 boxes of NERGIES will result in the largest possible profit, namely \$198.00. This program utilizes fully the available blending

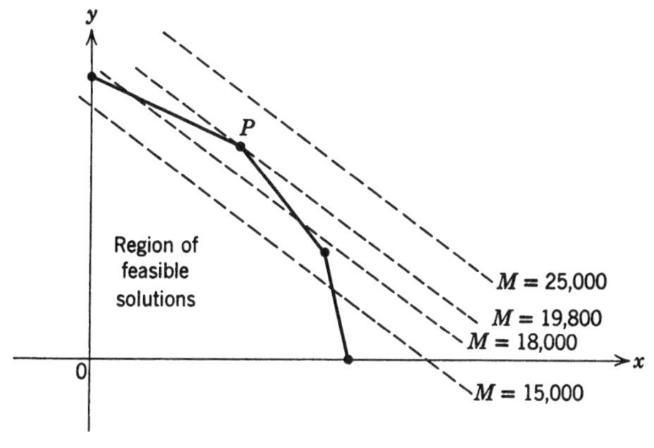

Figure 15

and cooking facilities, but leaves unused 4 of the 15 available hours of machine time in the packaging department, because $3(120) + 1(300)$ amounts to only 660 minutes, leaving 240 minutes of packaging machine time unutilized. This "left over" time constitutes the *slack* in this particular problem.

3 Alternate Optimal Solutions (Multiple Solutions)

In both examples of the preceding section, the optimal solution was *unique*. This means that any other feasible solution would turn out to be definitely inferior to the one obtained, i.e., would result in a greater cost or smaller profit, as the case may be.

Now, it is also possible to encounter linear programming problems where there is more than one, in fact *many* "equally good" optimal solutions. For instance, suppose we alter Example 2 above by reversing the unit profits, so that the confectioner makes 50¢ profit on each box of ERGIES and only 40¢ profit on each box of NERGIES. The profit formula is now expressed by

$$50x + 40y = M.$$

In this case, the family of parallel lines which corresponds to various values of the parameter M, appears as in Figure 16. The feasible

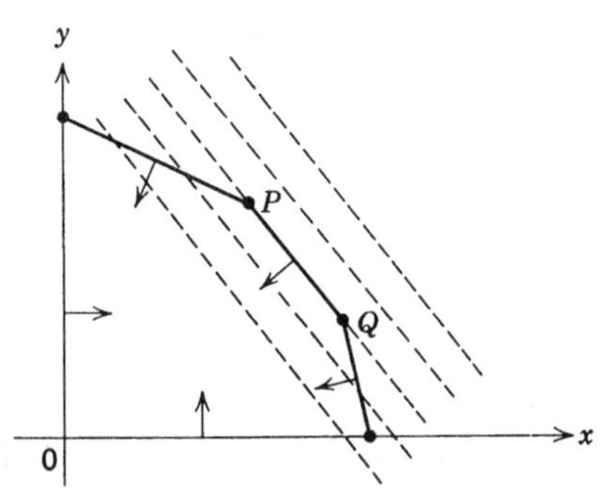

Figure 16

region is the same as in Figure 15, because we are assuming that the other conditions are the same as in that problem. One of the lines in the family now contains an entire segment $\overset{\bullet\!-\!\bullet}{PQ}$ of the boundary. This line has the equation

$$50x + 40y = 18{,}000$$

and corresponds to a profit value $M = \$180.00$. All points on the segment $\overset{\bullet\!-\!\bullet}{PQ}$ are feasible points which yield the same optimal value for M. For example, the coordinates of point P are

$$x = 120, \quad y = 300, \quad \text{giving } M = 50(120) + 40(300) = 18{,}000,$$

while the coordinates of point Q are

$$x = \frac{1800}{7} = 257\tfrac{1}{7}, \quad y = \frac{900}{7} = 128\tfrac{4}{7}, \text{ giving the same profit,}$$

$$\text{namely:} \quad M = 50\left(\frac{1800}{7}\right) + 40\left(\frac{900}{7}\right) = 18{,}000.$$

Of course, the latter solution involves fractional values for x and y. As a practical consideration, these values cannot be realized exactly by the confectioner.* Nevertheless, they are feasible values in the sense we are using this term, and represent an *alternate optimal solution.*

Still another alternate optimal solution, which is practical as well as feasible, is given by

$$x = 200, y = 200, \quad \text{giving} \quad M = 50(200) + 40(200) = 18{,}000.$$

This corresponds to a point *between* P and Q on the segment $\overset{\bullet\!-\!\bullet}{PQ}$. We invite the reader to find other alternate optimal feasible solutions, which can be realized in practice. Clearly, in a situation of this type, the confectioner has a choice among many "equally good" production programs and must use other criteria in making a decision as to which he will adopt.

We shall conclude this part of our discussion with one additional illustration of a simple linear programming problem in which both unique and multiple solutions arise.†

Example 3. The Broad Motor Company, manufacturer of the *Bulkswagon* (the latest word in compact cars for not-so-compact people), decides to sponsor a one-half hour television show featuring a comedian and a band. The Company insists that there must be at least 3 minutes of commercials. The T.V. network requires that the time allotted to commercials must not exceed 12 minutes, and under no circumstances may it exceed the time allotted to the comedian. The comedian is reluctant to work more than 20 minutes of the half-hour show, so the band is to be used to fill in any remaining time. The comedian costs the sponsor $150 per minute, the band $100 per minute and the commercials $50 per minute. Experience indicates that for every minute the comedian is on the air, 4000 additional viewers tune in; for every minute of band time, 2000 new viewers may be expected; but for every minute of commercial time,

* To be practical the solution must be a pair of integers, i.e., it must occur at a "lattice" point (x, y) where x and y are integers.

† This problem is adapted from Kemeny, Snell, and Thompson, *Finite Mathematics*, Chapter VI.

1000 people will tune *out*! How shall the available time be allotted if the sponsor is interested in:

a. obtaining the *maximum number of viewers*;

b. producing the show at *minimum cost.*

Analysis and Solution

Let x = number of minutes allotted to the comedian

y = number of minutes allotted to commercials

$\therefore \quad 30 - x - y$ = number of minutes allotted to the band

The problem imposes the following constraints on the values of the variables x and y:

$$\left.\begin{aligned} x &\geq 0 \\ y &\geq 0 \\ 30 - x - y &\geq 0 \end{aligned}\right\}$$ because the time allotted to each portion of the program can not be negative.

$y \geq 3$ restriction imposed by the sponsor.

$$\left.\begin{aligned} y &\leq 12 \\ y &\leq x \end{aligned}\right\}$$ restrictions imposed by the TV network.

$x \leq 20$ restriction imposed by the comedian.

If we let n be the number of people viewing the show, then the value of n is determined by the following formula:

$$n = 4000x + 2000(30 - x - y) - 1000y$$

or

$$n = 1000(2x - 3y + 60).$$

If we let c denote the cost of the program to the sponsor, then the value of c is expressed by the formula

$$c = 150x + 100(30 - x - y) + 50y$$

or

$$c = 50(x - y + 60).$$

We want to determine values for x and y which shall either maximize n or minimize c.

It is convenient to denote $2x - 3y$ by M, and $x - y$ by m. This enables us to restate the problem in the typical form described previously:

Determine: $x \geq 0$ and $y \geq 0$

$$\text{so that:} \quad \begin{cases} x + y \leq 30 \\ \quad\; y \geq 3 \\ \quad\; y \leq 12 \\ x - y \geq 0 \\ x \quad\;\; \leq 20 \end{cases}$$

$$\text{and so that:} \quad \begin{cases} \text{either} \quad M = 2x - 3y \quad \text{is a MAXIMUM} \\ \text{or} \quad m = x - y \quad \text{is a MINIMUM.} \end{cases}$$

The constraints define a closed polygon of feasible solutions. The vertices of this polygon are readily determined as intersections of adjacent boundary lines. The coordinates of these vertices are indicated in Figure 17.

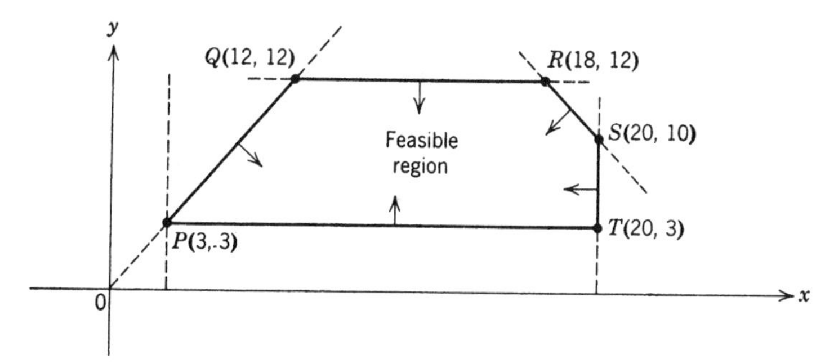

Figure 17

Treating M and m as parameters (i.e., assigning to each of these letters various values), we obtain two families of parallel lines, which we superimpose upon the diagram as indicated in Figure 18. A unique feasible maximum for M is attained at point $T = (20, 3)$. This corresponds to 91,000 viewers, as may be seen by substituting in the formula for n:

$$n = 1000(2 \cdot 20 - 3 \cdot 3 + 60) = 1000(91) = 91{,}000.$$

This is the maximum number of viewers, and it is achieved at a cost of \$3850, as may be seen by substituting in the formula for c:

$$c = 50(20 - 3 + 60) = 50(77) = 3850.$$

A lower cost can be achieved if the sponsor is willing to settle for fewer viewers. The minimum cost is attained at many feasible points, in fact all along the segment $\overleftrightarrow{PQ}$. The reader will verify, for example, that at point P,

$$x = 3, y = 3; \quad \therefore \ c = \$3000 \quad \text{and} \quad n = 57{,}000.$$

At point Q,

$$x = 12, y = 12; \quad \therefore \ c = \$3000 \quad \text{and} \quad n = 48{,}000.$$

The cost remains constant (\$3000) at all points along the segment $\overleftrightarrow{PQ}$, but the number of viewers varies along this segment from a maximum of 57,000 (at P) to a minimum of 48,000 (at Q). Among

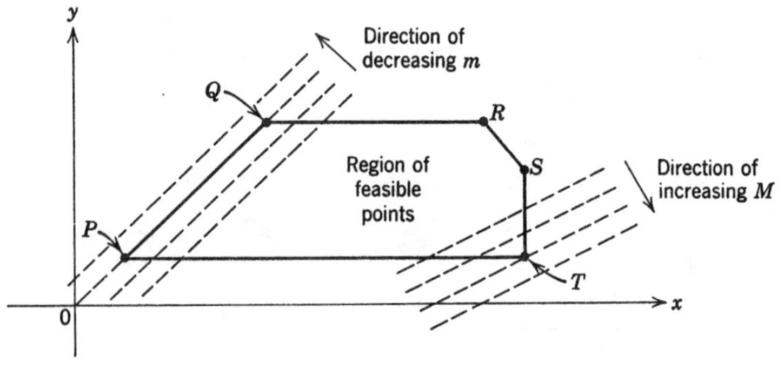

Figure 18

all these solutions, "equally good" from the cost point of view, the sponsor would undoubtedly prefer point P, because the values of x and y at this point bring in the most viewers for the same money.

In summary, if the Broad Motor Co. is interested in reaching the largest possible audience, the time should be allotted as follows: 20 minutes to the comedian, 3 minutes to commercials, and 7 minutes to the band. This solution coincides with the intuitive "common sense" approach since it allots the maximum permissible time to the comedian, who is the best audience "getter," while it cuts down to the minimum permissible time, the audience "loser," namely the commercials. On the other hand, if the sponsor is interested in saving money, his best program is not so intuitively obvious. The most expensive item is the comedian and the cheapest item, the commercials. However, a program entirely devoted to commercials is not feasible (and clearly undesirable). The maximum time allowed by the network is 12 minutes of commercials. If the sponsor adopts this in an effort to save money, he is forced to allot at least an equal amount of time to the comedian. Since this is the expensive item he will naturally keep this down to 12 minutes also. This solution corresponds to point Q, on our graph. Although it does achieve the goal of minimum cost, it is not his best program. By cutting both the comedian and the commercials down to 3 minutes each and offering his public 24 minutes of good music, he saves just as much money and keeps more of his audience. We trust that this will not only reaffirm the reader's belief in the worth of music, but will also convey to him some appreciation for the value of linear programming!

Exercises

1. Describe and sketch the set of points which corresponds graphically to each of the following constraints, or systems of constraints. (Do not assume non-negativity, unless this is implied by the constraints.)

(*a*) $x \leq 5$ (*b*) $x \geq 5$ (*c*) $\begin{cases} x \geq 5 \\ y \leq 2 \end{cases}$ (*d*) $\begin{cases} -1 \leq x \leq 6 \\ 0 \leq y \leq 4 \end{cases}$

(*e*) $\begin{cases} 0 \leq x \leq 10 \\ 0 \leq y \leq \ 8 \\ x + y \geq \ 6 \end{cases}$ (*f*) $\begin{cases} x \geq 0,\ y \geq 0 \\ x + 2y \geq 4 \\ x + 2y \leq 10 \\ y \leq x + 5 \\ x \leq y + 5 \end{cases}$

(*g*) $\begin{cases} 5x + y \geq -7 \\ -2x + 5y \leq 9 \\ 2x + 5y \leq 21 \\ 2x - y \leq 7 \\ x - 2y \leq 8 \\ x + 3y \geq -7 \end{cases}$ (*h*) $\begin{cases} 2x - y \geq -2 \\ 2x + y \geq 6 \\ x + 3y \geq 8 \\ x - y \leq 4 \end{cases}$

2. Write a set of constraints whose solution set corresponds to the *interior* of the triangle whose vertices are located at the three points whose coordinates are (0, 3), (2, 5) and (4, 1). How should the constraints be written so as to include the sides and vertices of the triangle?

3. In Exercise 1*h* above, the graph reveals that it would be redundant (superfluous) to stipulate $x \geq 0$ and $y \geq 0$. Prove this fact by *algebraic* manipulation of the inequalities, i.e., prove that the non-negativity of x and y are implied by the constraints of Exercise 1*h*.

(*Hint:* You should be able to prove an even "stronger" result, namely that the constraints actually imply $x \geq 1$ and $y \geq 1$.)

4. Mr. Hy P. Kondriak has been ordered by his physician to take daily at least 24 units of vitamin B_1 and at least 25 units of vitamin B_2. Unfortunately, these are not available in pure form, but the local drug stores sell HEALTH tablets at one cent each and STRENGTH capsules at three cents apiece. Each tablet contains 1 unit of B_1 and 5 units of B_2, while each capsule contains 4 units of B_1 and 1 unit of B_2.

Under these conditions, how many HEALTH tablets and how many STRENGTH capsules should Mr. Hy P. Kondriak purchase daily, in order to obtain the required vitamins at *minimum cost*?

5. The Broad Motor Company manufactures two basic car models, the luxurious *Pontillac* and the low cost *Bulkswagon*. These are sold to car dealers at a profit of $200 per Pontillac and $100 per Bulkswagon. A Pontillac requires, on the average, 150 man-hours for assembly, 50 man-hours for painting and finishing, and 10 man-hours for checking out and testing. The Bulkswagon averages 60 man-hours for assembly, 40 man-hours for painting and finishing, and 20 man-hours for checkout and testing. During each production run, there are 30,000 man-hours available in the assembly shops, 13,000 man-hours in the painting and finishing shops, and 5,000 man-hours in the checking and testing division.

How many Pontillacs and how many Bulkswagons should the Broad Motor Company plan to produce, in order to realize the *greatest possible profit* from each production run?

(*Answer:* 140 Pontillacs, 150 Bulkswagons for a maximum profit of $33,000.)

6. (*a*) In the preceding problem (Exercise 5), what would be the most profitable production program if both Pontillacs and Bulkswagons were sold to the dealers at the same profit of $100 per car?

(*Answer:* 100 Pontillacs, 200 Bulkswagons for a maximum profit of $30,000.)

(*b*) Suppose that the demand for luxury cars were to increase to the point where the Company could realize three times as much profit on a Pontillac as on a Bulkswagon. Show that it would then no longer pay to produce the smaller car at all. In that case, how many could the company produce?

(*Answer:* 200 Pontillacs, 0 Bulkswagons.)

(*c*) If the public preferred the smaller car to such an extent that the company could make more money on Bulkswagons than on Pontillacs, at what point would it pay to switch all of its production facilities over to sole production of the smaller car?

(*Answer:* This would occur if the company could realize at least twice as much profit on Bulkswagons. At this point there are alternate optimal feasible programs available to the company. Beyond this point it pays to make only Bulkswagons.)

7. (*a*) In Example 3 (see Section 3), calculate the value of M, i.e., the value of $2x - 3y$, at each of the vertices R and S (see Figure 17). Determine the midpoint of segment $\overline{RS}$, and calculate the value of M at this midpoint. Compare these three values of M. At which point is it largest; at which smallest? Is there any relation between the value at the midpoint and the value at the ends?

(*b*) Compare the value of M at point R with its value at point Q (Figure 17). Now determine the midpoint of segment $\overset{\bullet\!-\!\bullet}{QR}$, and compare the value of M at this midpoint with the values of M at each end of the segment. Also find several other points on segment $\overset{\bullet\!-\!\bullet}{QR}$, and compare the values which M assumes at these points with the values already obtained. Formulate a possible generalization.

8. Generalize the previous problem by proving that if $M = ax + by$, then the value of M at the midpoint of any segment is equal to the *average* of its values at the ends of the segment. Can the value of M at points within a segment ever exceed *both* of the values it assumes at the ends of the segment? Does your answer to the last question offer any clue as to the behavior of a linear form $M = ax + by$, at points inside a closed polygon, as compared with its behavior on the boundary? (We shall discuss this question in the next chapter.)

PART TWO

Convex Sets in the Cartesian Plane and the Fundamental Extreme Point Theorem

1. Preliminary Remarks

In part one, we introduced several simple linear programming problems and observed that in each case the feasible solutions were conveniently represented by certain sets of points in a plane. These geometric notions (*point* and *plane*) were useful for the purpose of obtaining a visual picture of these solutions, but we should not forget that the solutions themselves were actually (pairs of) real numbers. Other geometric configurations which arose quite naturally in our treatment of linear programming problems were: *line*, *segment*, *polygon*, *half plane*, etc., all introduced in a more or less intuitive fashion.

Geometric intuition is, of course, a very fine thing, and the reader will do well to develop his own as fully as he possibly can. But a true mathematician prefers, wherever possible, to crown his intuitive insights with a sound logical analysis, bringing them together into one carefully constructed theory. In the present case, this can be accomplished in a fairly elegant way which, as often happens, yields fresh insights into the problem or phenomenon under scrutiny.

Such a mathematical theory usually turns out to be more or less abstract and may require a certain degree of sophistication or "mathematical maturity" on the part of the reader. We shall keep these demands to a minimum, by confining our sets of points to two dimensions. However, much of the discussion readily generalizes to higher dimensional spaces. If the reader is interested in such a more general analysis he can find a number of good treatments among the references (e.g., References 2, 3, or 8).

2. The Linear Relation in the Cartesian Plane

Let us begin by defining a *point* to be an ordered pair of real numbers. For example, (2, 3), (−4, 3/5), (0, 0), etc. are all points. If (x, y) denotes a point P, the value of x is called the *first coordinate* of P, and the value of y is called the *second coordinate* of P.

The set of all ordered pairs of real numbers is called the *cartesian plane*. Denoting by R, the set of all real numbers, it is customary to denote the cartesian plane by $R \times R$ or R^2. Using the language and notation of sets, this may also be expressed as follows:

$$R^2 = R \times R = \{ (x, y) \mid x \in R \text{ and } y \in R \}.$$

Any subset of $R \times R$ is called a *relation*. For example, the set of all points whose first coordinate is 0 is a relation. This particular relation is the set

$$Y = \{ (x, y) \mid x = 0 \text{ and } y \in R \},$$

and this set is called the *y-axis*. It is usually specified by simply writing the equation $x = 0$. Similarly, the set of all points whose second coordinate is 0 is also a relation. It consists of the set,

$$X = \{ (x, y) \mid y = 0 \text{ and } x \in R \}$$

and is called the *x-axis*. Its defining equation is simply $y = 0$.

Both of these sets are special examples of a *linear relation.* We define this in the following way:

DEFINITION 1. A *linear relation* is a set of points in R^2 whose coordinates x and y satisfy a linear equation: $ax + by = c$, where a, b, c are real numbers with a, b *not both* 0.

Let us examine, briefly, the reason why this type of relation is called *linear.* Evidently the terminology is intended to convey the idea that this relation has something to do with the intuitive geometric notion of a line. Now, our cartesian plane, as we have defined

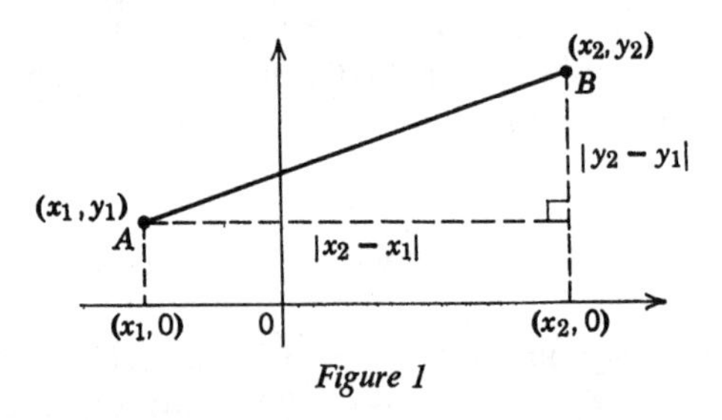

Figure 1

it thus far, is merely an aggregation of ordered pairs of real numbers and we are calling these ordered pairs *points.* To satisfy our geometric intuition we must impose a geometric "structure" on our cartesian plane. We shall do this by defining a *distance* between any two points of R^2. Our definition is motivated by the Pythagorean Theorem of Euclidean Plane Geometry.

DEFINITION 2. The distance AB between the points $A = (x_1, y_1)$ and $B = (x_2, y_2)$ is defined as

$$AB = \sqrt{(x_2 - x_1)^2 + (y_2 - y_1)^2}.$$

Figure 1 indicates the close connection between this definition and the Pythagorean Theorem. But the definition itself is purely "analytic," and it enables us to study the geometric concept of distance by referring back to familiar properties of the real numbers rather than by relying upon the appearance of a diagram. In the present instance, we shall use our definition of distance to show that the set of points which make up a linear relation actually does possess a property characteristic of the straight line in an intuitive sense. This property is embodied in the following theorem:

THEOREM 1. If A, B, and C are any three points in a linear relation, then one of the distances AB, BC, and AC is always equal to the sum of the other two.

Proof

Suppose l is a linear relation.

$$l = \{ (x, y) \in R^2 \mid ax + by = c \}$$

where a, b, and c are real numbers with a and b not both zero. Consider any three points A, B, and C, of l, where

$$A = (x_1, y_1), \qquad B = (x_2, y_2), \qquad C = (x_3, y_3).$$

We may suppose A, B, and C are *distinct*, for if not, say $A = B$,

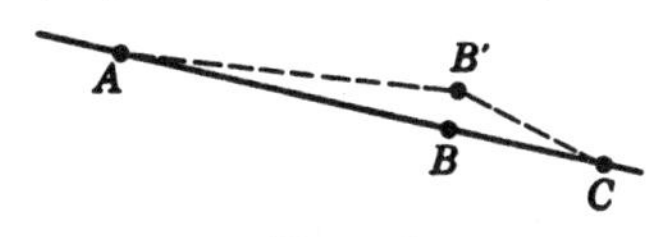

Figure 2

then $AB = 0$, $AC = BC$, and the equation $AB + BC = AC$ is trivially true in such a case. Now we know that

$$ax_1 + by_1 = c$$
$$ax_2 + by_2 = c$$
$$ax_3 + by_3 = c.$$

If we subtract the first equation from the second, the second from the third, and the first from the third, we obtain

$$a(x_2 - x_1) + b(y_2 - y_1) = 0$$
$$a(x_3 - x_2) + b(y_3 - y_2) = 0$$
$$a(x_3 - x_1) + b(y_3 - y_1) = 0.$$

We know that a and b are not both 0. Let us suppose that $a \neq 0$ (the argument is completely similar if we suppose $b \neq 0$). It follows that

$$(x_2 - x_1) = -\frac{b}{a}(y_2 - y_1)$$

$$(x_3 - x_2) = -\frac{b}{a}(y_3 - y_2)$$

$$(x_3 - x_1) = -\frac{b}{a}(y_3 - y_1).$$

We now compute the distances AB, BC, and AC.

$$AB = \sqrt{(x_2 - x_1)^2 + (y_2 - y_1)^2} = \sqrt{\frac{b^2}{a^2}(y_2 - y_1)^2 + (y_2 - y_1)^2}$$

$$= \sqrt{\left(\frac{b^2}{a^2} + 1\right)(y_2 - y_1)^2} = \sqrt{\frac{a^2 + b^2}{a^2}}\sqrt{(y_2 - y_1)^2}$$

$$\therefore\ AB = \frac{\sqrt{a^2 + b^2}}{|a|}|y_2 - y_1| \quad \text{(note that } \sqrt{r^2} = |r|, \text{ for all real } r).*$$

Similarly,

$$BC = \frac{\sqrt{a^2 + b^2}}{|a|}|y_3 - y_2| \quad \text{and} \quad AC = \frac{\sqrt{a^2 + b^2}}{|a|}|y_3 - y_1|.$$

We may suppose the notation A, B, C so chosen that

$$y_1 \leq y_2 \leq y_3,$$

in which case we may write

$$|y_2 - y_1| = y_2 - y_1, \quad |y_3 - y_2| = y_3 - y_2, \quad |y_3 - y_1| = y_3 - y_1.$$

It immediately follows that

$$AB + BC = \frac{\sqrt{a^2 + b^2}}{|a|}[(y_2 - y_1) + (y_3 - y_2)] = \frac{\sqrt{a^2 + b^2}}{|a|}(y_3 - y_1),$$

i.e.,

$$AB + BC = AC.$$

This completes our proof of Theorem 1.

Having verified that a linear relation l, meets an important intuitive requirement for being a straight line, let us now investigate the properties of its complementary set l'. This is defined as the set of points remaining in R^2 after removal of the points of l. Formally, we may define l' as follows:

$$l' = \{(x, y) \in R^2 \mid ax + by \neq c\}. \qquad (a \text{ or } b \neq 0)$$

* The symbol "$|r|$" is called "the absolute value of r." Its value is always ≥ 0. Conventionally, $\sqrt{r^2} \geq 0$ for all real r.

The points of l' fall naturally into one or the other of two disjoint (completely distinct) classes.

$$p = \{(x, y) \in R^2 \mid ax + by > c\}$$
$$q = \{(x, y) \in R^2 \mid ax + by < c\}. \qquad (a \text{ or } b \neq 0)$$

Every point in R^2 is evidently a member of exactly one of the three sets l, p, or q, which are therefore said to "partition" R^2.

The sets p and q are called *open half planes.*

The set l is called the *boundary* of each of these half planes. If the boundary l is united with the open half plane p, the resulting set is called a *closed half plane* and is denoted by $\bar{p}$. In the language of sets we call this the *union* of l and p and write

$$\bar{p} = l \cup p.$$

The set $\bar{p}$ may also be described as follows:

$$\bar{p} = \{(x, y) \in R^2 \mid ax + by \geq c\}. \qquad (a \text{ or } b \neq 0)$$

Similarly, for the closed half plane $\bar{q}$:

$$\bar{q} = l \cup q = \{(x, y) \in R^2 \mid ax + by \leq c\}. \qquad (a \text{ or } b \neq 0)$$

Observe that an open half plane is defined by a *strict* linear inequality, such as

$$ax + by > c \quad \text{or} \quad ax + by < c.$$

Either of these inequalities can be changed into the other form by simply multiplying both members by -1. A closed half plane is defined by a *weak* linear inequality, such as

$$ax + by \geq c \quad \text{or} \quad ax + by \leq c.$$

A linear *equality*, on the other hand, defines a straight line. It is interesting to note that we can regard the notion of a closed half plane as basic, and define a straight line as the intersection of two closed half planes of the type $\bar{p}$ and $\bar{q}$ described above.

In the following exercises, linear relations are shown to possess still further properties, ordinarily associated with straight lines. For this reason we feel free to use the term "straight line" synonymously with "linear relation."

Exercises

1. Prove that a straight line (i.e., linear relation) is determined by any two of its (distinct) points. Do this by proving that if (x, y_1) and (x_2, y_2) are distinct points of l, then every point (x, y) of l satisfies the equation

$$(y_2 - y_1)x - (x_2 - x_1)y = x_1y_2 - x_2y_1.$$

2. If $x_1 \neq x_2$ in Exercise 1 prove that every point of l satisfies the equation

$$y = mx + b,$$

where $m = \dfrac{y_2 - y_1}{x_2 - x_1}$ (this is called the *slope* of l),

and $b = \dfrac{x_2 y_1 - y_2 x_1}{x_2 - x_1}$ (this is called the *y-intercept* of l).

3. Prove that the equation $y = mx + b$ derived in Exercise 2 is *equivalent* to the equation

$$y - y_1 = m(x - x_1), \quad \text{where} \quad m = \frac{y_2 - y_1}{x_2 - x_1}.$$

(*Note: Equivalent equations* mean equations which define the same solution sets in $R \times R$.)

4. Prove that the slope $m = (y_2 - y_1)/(x_2 - x_1)$ does not depend upon which pair of points of l have been selected as (x_1, y_1) and (x_2, y_2), so long as these are distinct points of l, whose first coordinates are distinct.

5. If l is defined by $ax + by = c$, prove that l has a slope m, if and only if $b \neq 0$, in which case $m = \dfrac{a}{b}$.

6. Prove that if the (distinct) lines l and l' have slopes m and m' respectively, then l and l' fail to intersect, if and only if, $m = m'$.

(*Note:* Two lines in R which fail to intersect are called *parallel*.)

7. If l_1 and l_2 are distinct lines defined respectively by,

$$a_1x + b_1y = c_1 \quad \text{and} \quad a_2x + b_2y = c_2,$$

then l_1 is parallel to l_2, if and only if,

$$a_1b_2 - a_2b_1 = 0.$$

(*Note:* This theorem is a bit more general than the one proved in Exercise 6, because it does not require that the lines l and l' have slopes.)

8. If the lines l_1 and l_2 of Exercise 7 intersect at point (x, y), prove that

$$x = \frac{c_1b_2 - c_2b_1}{a_1b_2 - a_2b_1} \quad \text{and} \quad y = \frac{a_1c_2 - a_2c_1}{a_1b_2 - a_2b_1}.$$

3. Convex Sets in the Cartesian Plane

Straight lines and half planes in R^2 possess an important property called *convexity*. The notion of a convex set plays such a prominent role in linear programming (and other areas of modern applied mathematics) that we shall devote some time to it.

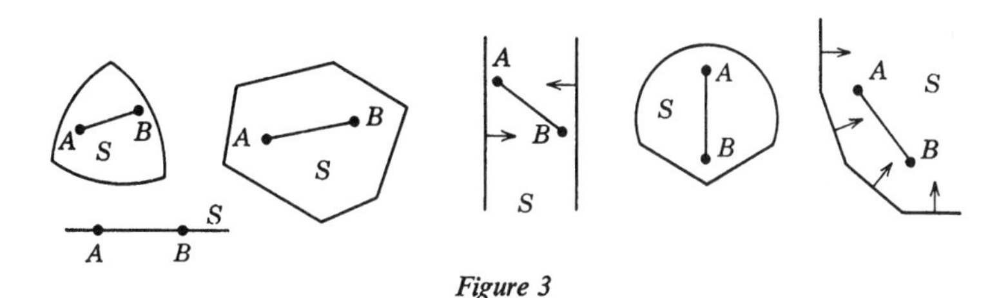

Figure 3

In geometric language, a set of points S, is said to be *convex*, if and only if, the set S contains the entire *segment* joining any two of its points. The sets depicted in Figure 3 are all convex sets.

The sets depicted in Figure 4 are not convex sets.

Since the idea of convexity depends on the notion of *segment*, we should try to make the latter concept precise. Intuitively, a segment is a portion of a line containing all those points of the line which are *between* two given points. We may possibly also wish to

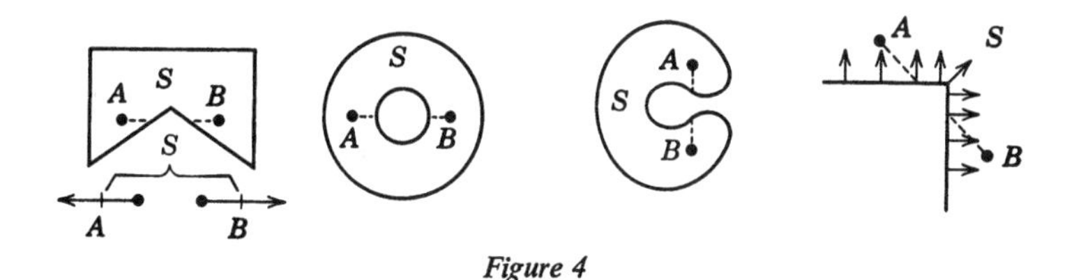

Figure 4

include in the segment one or both of its end points. Our problem is to express these ideas algebraically.

In order to accomplish this task, it is convenient to introduce an analytic description of a segment. This can be done by generalizing the well known *midpoint formula* of plane coordinate geometry. Consider two distinct points: $A = (x_1, y_1)$ and $B = (x_2, y_2)$. If $M = (x, y)$ is the midpoint of segment $\overset{\bullet\!-\!\bullet}{AB}$, then the coordinates of

M are given by $x = \frac{1}{2}x_1 + \frac{1}{2}x_2$ and $y = \frac{1}{2}y_1 + \frac{1}{2}y_2$. More generally, if M divides $\overleftrightarrow{AB}$ in the ratio $p : q$, i.e., if $\frac{AM}{MB} = \frac{p}{q}$, then by using similar triangles in Figure 5, it is readily seen that

$$\frac{x - x_1}{x_2 - x} = \frac{p}{q} = \frac{y - y_1}{y_2 - y}.$$

Solving the first of these equations for x yields:

$$x = \frac{q}{p+q}\,x_1 + \frac{p}{p+q}\,x_2.$$

Similarly, solving for y:

$$y = \frac{q}{p+q}\,y_1 + \frac{p}{p+q}\,y_2.$$

(The special case $p = q$, yields the midpoint formula.)

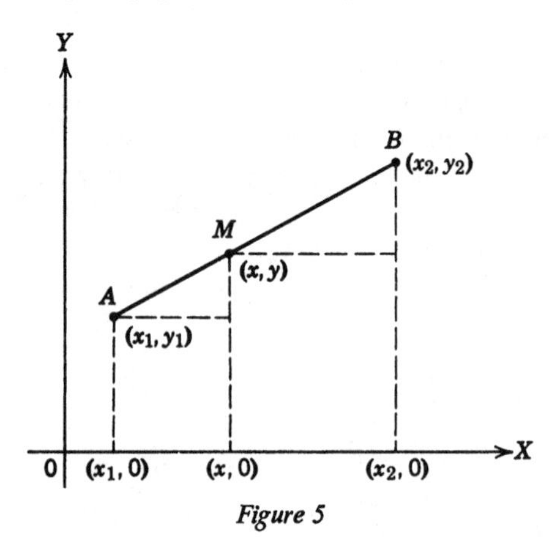

Figure 5

Now, if we let $t = \frac{p}{p+q}$, then $1 - t = \frac{q}{p+q}$ and the coordinates of the point M are then expressed as follows:

$$x = (1 - t)x_1 + tx_2 \quad \text{and} \quad y = (1 - t)y_1 + ty_2.$$

Here t is restricted by definition so that $0 < t < 1$. As t varies between 0 and 1, point M varies over the segment $\overline{AB}$. On the other hand, if we remove the restriction and allow t to vary over all real numbers, then we shall obtain the entire line l determined by the two points A and B. To see this let us start with a line l, defined by

$$ax + by = c \qquad \text{(where } a \text{ and } b \text{ are not both 0).}$$

Let $A = (x_1, y_1)$ be a fixed point of l, so that

$$ax_1 + by_1 = c.$$

Then, for any point (x, y) of l, we have

$$a(x - x_1) + b(y - y_1) = 0.$$

Now let r be *any* real number and consider all those points (x, y) for which

$$x - x_1 = rb \quad \text{and} \quad y - y_1 = -ra.$$

These points are obviously points of l, because for each of them

$$a(x - x_1) + b(y - y_1) = a(rb) + b(-ra) = 0.$$

Conversely, if (x, y) is any point of l, then we can always find a real value r, such that $x - x_1 = rb$ and $y - y_1 = -ra$. This is possible because a and b are not both 0. Suppose, for example, that $a \neq 0$. Then we need merely choose

$$r = \frac{y - y_1}{-a}.$$

This choice of r, automatically makes $y - y_1 = -ra$, from which it follows that

$$a(x - x_1) + b(y - y_1) = a(x - x_1) + b(-ra) = 0$$

$$\therefore \quad a(x - x_1) = a(br)$$

$$\therefore \quad (x - x_1) = br \qquad \text{(since } a \neq 0\text{).}$$

If $a = 0$, then $b \neq 0$ and the argument is quite similar.

We have thus shown that the set of points l, defined by the *single* linear equation

$$\text{(1)} \qquad ax + by = c \qquad \text{(with } a \text{ and } b \text{ not both 0)}$$

is identical with the set of points defined by the *pair* of equations

$$\text{(2)} \qquad x = x_1 + rb \quad \text{and} \quad y = y_1 + r(-a), \ (r \in R)$$

where (x_1, y_1) is any (fixed) point of l. The real variable r appearing in this pair of equations is called a *parameter* and the pair of equations is called a *parametric representation* of line l.

In a parametric representation such as (2), the parameter r may assume any real value. When $r = 0$, one obtains $x = x_1$ and $y = y_1$. These are the coordinates of one particular point A of l. If $B = (x_2, y_2)$ is another particular point of l, there must be some value of r, say $r = k$, for which

$$x_2 = x_1 + kb \quad \text{and} \quad y_2 = y_1 + k(-a).$$

Furthermore, if B is distinct from A, then $k \neq 0$. It then follows that $b = (x_2 - x_1)/k$ and $(-a) = (y_2 - y_1)/k$. These expressions for b and $(-a)$ may now be substituted in (2):

$$x = x_1 + \frac{r}{k}(x_2 - x_1) \quad \text{and} \quad y = y_1 + \frac{r}{k}(y_2 - y_1).$$

Since r can be any real number, we may denote the ratio r/k by t, where t can now be any real number also. Our parametric representation then takes on the following useful form:

(3) $\quad x = x_1 + t(x_2 - x_1) \quad \text{and} \quad y = y_1 + t(y_2 - y_1), \quad (t \in R).$

This is also conveniently rewritten,

(4) $\quad x = (1 - t)x_1 + tx_2 \quad \text{and} \quad y = (1 - t)y_1 + ty_2, \quad (t \in R).$

These are the same expressions which we obtained from geometric considerations previously.

Observe that

$$\text{when } t = 0, \text{ then } x = x_1 \text{ and } y = y_1;$$

$$\text{when } t = 1, \text{ then } x = x_2 \text{ and } y = y_2.$$

The two special parameter values 0 and 1, therefore, yield respectively the particular points A and B. Moreover, we now also observe that if t assumes a real value between 0 and 1, more precisely if $0 < t < 1$, then the resulting point C must "lie between" points A and B. As the latter idea of "betweenness" for points is essentially an intuitive notion, we shall try to make the last assertion more precise. We accomplish this with the aid of the following very useful theorem:

THEOREM 2. If C is any point on line l, with parametric representation given either by (3) or (4), and if C corresponds to the parameter value t, then $AC = |t|\, AB$.

Proof

If C is the point (x, y), then by (3) we have

$$x - x_1 = t(x_2 - x_1) \quad \text{and} \quad y - y_1 = t(y_2 - y_1).$$

Therefore,

$$\begin{aligned} AC &= \sqrt{(x - x_1)^2 + (y - y_1)^2} \\ &= \sqrt{t^2(x_2 - x_1)^2 + t^2(y_2 - y_1)^2} \\ &= \sqrt{t^2}\sqrt{(x_2 - x_1)^2 + (y_2 - y_1)^2} \\ AC &= |t|\, AB. \qquad \text{(Q.E.D.)} \end{aligned}$$

Suppose now, that $0 < t < 1$. Then, $0 < 1 - t < 1$. Applying Theorem 2 directly, gives $AC = t\,AB$. On the other hand we can rewrite the parametric representation (4) as follows:

$$(4)' \qquad x = tx_2 + (1 - t)x_1 \quad \text{and} \quad y = ty_2 + (1 - t)y_1, \ (t \in R),$$

and then apply Theorem 2 with the roles of A and B interchanged and $(1 - t)$ replacing t. This gives $BC = (1 - t)BA$, which can also be written $CB = (1 - t)AB$. If we add AC and CB, we get

$$AC + CB = t\,AB + (1 - t)AB = AB.$$

This equality gives us a precise mathematical way of stating that point C lies between points A and B.

DEFINITION 3. Point C is *between* points A and B, if and only if, $AC + CB = AB$.

We can now also give a precise definition of the term *segment.*

DEFINITION 4. If $A = (x_1, y_1)$ and $B = (x_2, y_2)$ are points in R^2, then we define

(*a*) The *closed segment* $\overset{\bullet\!-\!\bullet}{AB}$ is the set consisting of points A and B, together with all points between A and B.

(*b*) The *open segment* $\overline{AB}$ is the set consisting of points *between* A and B. Using a parametric representation, we may express the closed segment $\overset{\bullet\!-\!\bullet}{AB}$ as follows:

$$\begin{aligned} \overset{\bullet\!-\!\bullet}{AB} = \{(x, y) \in R^2 \mid x &= (1 - t)x_1 + tx_2 \quad \text{and} \\ y &= (1 - t)y_1 + ty_2, \quad \text{for} \quad 0 \le t \le 1\}. \end{aligned}$$

The open segment $\overline{AB}$ has the same parametric representation except that the parameter t must now be restricted to values between 0 and 1, i.e., $0 < t < 1$. Either endpoint can be included in or omitted from the segment by a slight alteration in the inequality which specifies the permissible values of the parameter t. Thus,

$0 \leq t < 1$ includes point A but not B (notation: $\overline{\overset{\bullet}{A}B}$);

$0 < t \leq 1$ includes point B but not A (notation: $\overline{A\overset{\bullet}{B}}$).

For later use it will be convenient to generalize our parametric representations somewhat, by use of the following theorem:

THEOREM 3. Let a parametric representation be chosen for a line l and let P, Q, and R be three distinct points of l with corresponding parameter values p, q, and r, respectively. If $p < q < r$, then point Q is between P and R.

Proof

This can be proved directly, using Definition 3. A less tedious procedure is to define a new parameter t' as follows:

$$t' = \frac{t - p}{r - p}.$$

Then points P and R will correspond respectively to the new parameter values $t' = 0$ and $t' = 1$, while point Q will correspond to the new parameter value $(q - p)/(r - p)$. We leave it to the reader to show that this last value lies between 0 and 1.

For later use, we also define the terms *ray* and *half line.*

DEFINITION 5. Let the point P, on line l, correspond to the parameter value p.

(*a*) Each of the sets defined by $t \leq p$ or $t \geq p$ is called a *ray.* P is called the *endpoint* of the ray.

(*b*) Each of the sets defined by $t < p$ or $t > p$ is called a *half line.* P is called the *endpoint* of the half line.

Notice that a ray contains its endpoint, while a half line does not.

Our intuitive description of a convex set (see page 29 above) now acquires the status of a fairly precise definition.

DEFINITION 6. A set of points is called *convex*, if and only if, whenever it contains points A and B, then it also contains the segment $\overline{AB}$ (i.e., it contains A, B and all points between A and B).

We are now in a position to establish several theorems which are of considerable importance for linear programming.

THEOREM 4. Any half plane (closed or open) is a convex set.

Proof

Before we give a formal proof, we remark that this theorem is *intuitively* obvious! A half plane consists of all points "on one side" of a line. If we select two points "on the same side" of a line, it seems quite clear that the entire segment between them must also lie on this same side. Just look at the diagram!

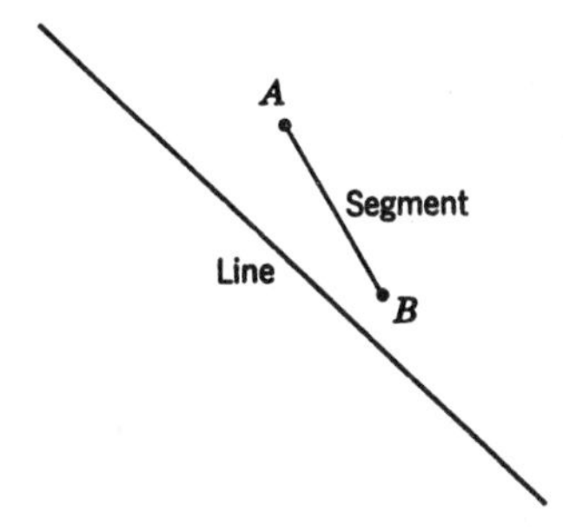

Nevertheless, a mathematical proof endeavors to go beyond a mere intuitive feeling, however strong it may be. In the present instance, the proof is entirely *analytic* (i.e., algebraic) and in no way dependent upon a diagram.

We shall prove the theorem for closed half planes. The proof for open half planes is practically the same.

If (x_1, y_1) and (x_2, y_2) are points of a closed half plane defined by $ax + by \leq c$, then

$$ax_1 + by_1 \leq c \quad \text{and} \quad ax_2 + by_2 \leq c.$$

Now, if (x', y') is any point on the (closed) segment determined by these points, then, by Definition 4

$$x' = (1 - t)x_1 + tx_2 \quad \text{and} \quad y' = (1 - t)y_1 + ty_2, \ (0 \leq t \leq 1).$$

Our problem is to show that (x', y') also belongs to the closed half plane, that is, we must show that $ax' + by' \leq c$. This result is immediately seen to be true by direct substitution.

$$\begin{aligned} ax' + by' &= a(1 - t)x_1 + atx_2 + b(1 - t)y_1 + bty_2 \\ &= (1 - t)(ax_1 + by_1) + t(ax_2 + by_2) \\ &\leq (1 - t)c + tc = c - tc + tc \end{aligned}$$

$$ax' + by' \leq c. \qquad \text{(Q.E.D.)}$$

THEOREM 5. The intersection of any collection of convex sets is a convex set.

Proof

Consider any collection of convex sets (finite or infinite in number). If their intersection contains points A and B, then these points must be contained in *each* convex set of the given collection (because the intersection of a collection of sets contains all points common to these sets). Hence, *each* of these convex sets contains the segment AB. Therefore, the segment $\overline{AB}$ is contained in the intersection of these convex sets. The intersection is therefore a convex set, by Definition 6.

An immediate consequence of this theorem is:

THEOREM 6. The intersection of any set of half planes is a convex set.

Exercises

1. Prove that any straight line in R^2 is a convex set.
2. *a.* If a set contains exactly one point, is it a convex set? Explain.
   b. Is the null set (empty set) a convex set? Explain.
   c. If a set contains a finite number of distinct points, is it a convex set? Explain.
3. If a and b are real numbers such that $a < b$, prove that the set of all points (x, y) for which $a \leq x \leq b$, is a convex set (see Figure 6).

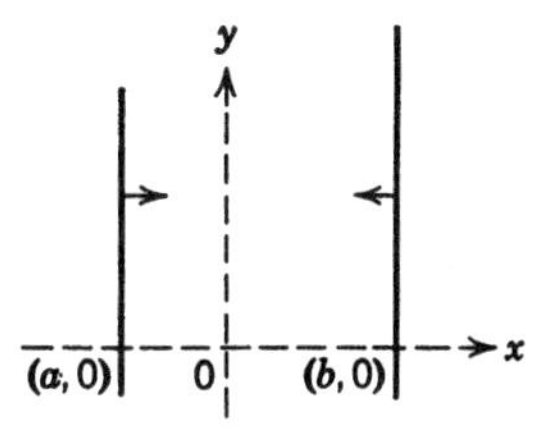

Figure 6

4. If a, b, c, and d, are real numbers such that $a < b$ and $c < d$, prove that the set of all points (x, y) for which

$$a \leq x \leq b \quad \text{and} \quad c \leq y \leq d$$

is a convex set. (Note: this set of points is a "rectangle," see Figure 7.)

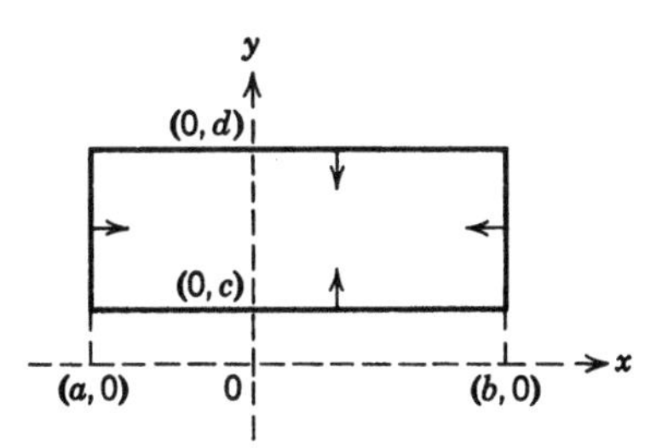

Figure 7

5. Prove that a *circular disk* is a convex set. Do this by first defining a circular disk as a set of points (x, y), such that

$$(x - h)^2 + (y - k)^2 \leq a^2$$

where h, k, and a are fixed real numbers (see Figure 8).

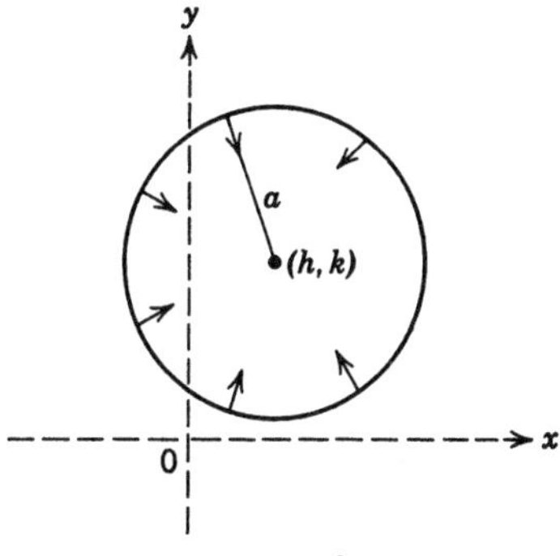

Figure 8

4. Convex Subsets of a Line

It is possible to classify completely, all the possible convex subsets of a line. This will be quite valuable in our further study of the convex subsets of the cartesian plane.

Let k be any convex subset of a line l in R^2. Of course l is itself a convex set (see Exercise 1 preceding this section). If $k \neq l$, there must exist a point D in l, which is not in k. Choose any parametric representation for l, and let D correspond to the parameter value d. Let A be any point in k and let its corresponding parameter value be a. Since A and D are distinct points, $a \neq d$. Therefore, either $a < d$ or $d < a$.

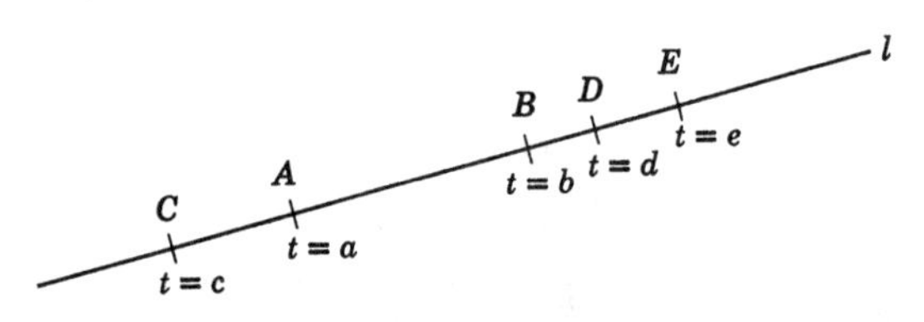

Suppose $a < d$ (a similar argument holds when $d < a$). Then all points of k must correspond to parameter values which are also less than d. For, if there were a point E in k with parameter value $e \geq d$, then because k is convex, every point in segment $\overline{AE}$, including D(!), would belong to k. This contradicts the fact that D was chosen *not* in k. We see therefore, that if $a < d$, then the set of all parameter values corresponding to points of k, has an *upper bound.* (The real number d is such an upper bound in this case.)

Now, there is a well known property of the real number system, called *completeness*, which asserts that if a set of real numbers has an upper bound, then there exists among all its upper bounds a smallest one. This is called the *least upper bound.* In the present case, let this least upper bound be the real number b. Using b as a parameter value, we obtain a point B on line l, having the property that every point of l between A and B belongs to k, but every point of l "beyond" B, i.e., whose parameter value is *larger* than b, does not belong to k. The point B itself, may or may not belong to k.

If the points of l which lie "beyond" B are the *only* points of l

which are not in k, then k is clearly the *ray* defined by the parameter values $t \leq b$ (see Definition 5*a*). If, in addition, B does not belong to k, then k consists of the *half line* defined by $t < b$ (see Definition 5b). In either case B is an endpoint of the ray or half line k.

If, on the other hand, there are still points of l, "below" B, i.e., with parameter values *less* than b, which do not belong to k, then these points must lie "beyond" A, i.e., they must correspond to parameter values which are actually *less* than a. (This is so because A and all points between A and B are members of k.) Once again, using the completeness property of the real number system, we see that there is a least upper bound c, for the parameter values of these remaining points of l which are not in k and are beyond A. Let point C on l correspond to parameter value c. All points of l between C and B, now clearly belong to k. C itself, may or may not belong to k. No point with parameter value less than c can belong to k, as k is convex and would contain all points between this point and point A, including C itself, thus yielding smaller upper bounds than c, for the set whose least upper bound was supposed to be c.

In short, k can now only consist of all points between C and B, and may or may not include either endpoint B or C. This makes k a segment of l, with or without either endpoint. All of this is now briefly summarized in the following theorem:

THEOREM 7. A (nonempty) convex subset of a line l must be one and only one of the following: (*a*) the *entire line* l; (*b*) a *ray* on l; (*c*) a *half line* on l; (*d*) a *segment* on l (with or without either endpoint).

Note that a single point of l is a convex subset and may be regarded as a special case of a segment (see Exercise 2*a* of the previous section). Except for this trivial case the remaining convex subsets of a line may be referred to as "one-dimensional" convex sets. In the next section we shall study "two-dimensional" convex sets. These are far more complicated and we shall concentrate our attention on a special kind known as "polygonal convex sets" which are of great importance in linear programming.

Exercises

1. Prove Theorem 7 for the case where $d < a$. (Observe that this requires the fact that every set of real numbers, which has a lower bound, has a greatest lower bound. Can you derive this fact from the corresponding assertion about the existence of a least upper bound?)

2. Prove that any point P of a line l, "separates" l into two half lines with the property that a segment joining a point of one of these half lines to a point of the other half line, always contains the point P.

3. Given points A and B on line l, let us call points A and B "boundary" points of the closed segment $\overset{\bullet-\bullet}{AB}$; call all points which are between A and B, "interior" points of $\overset{\bullet-\bullet}{AB}$; call all other points of line l, "exterior" points of $\overset{\bullet-\bullet}{AB}$. Prove that any segment joining an interior point to an exterior point, "crosses the boundary."

5. Polygonal Convex Sets and Linear Forms

In Section 3, we proved that the intersection of any set of half planes is a convex set. In particular, if all these half planes are closed and if there are only a finite number of them, we shall call their intersection a *polygonal* convex set.

DEFINITION 7. The intersection of a finite number (≥ 1) of closed half planes is called a *polygonal convex set.*

We encountered a number of examples of polygonal convex sets in Part One (see Figures 11, 14, and 17 of that Part). Any closed half plane qualifies as a polygonal convex set and so does any straight line or even any closed segment. (Verify that these are all expressible as intersections of a finite number of half planes.) Our chief interest centers, however, upon convex *polygons* and *unbounded* convex polygonal *regions.* We shall define these sets in due course.

According to Definition 7, a polygonal convex set $\mathscr{K}$ consists of all points (x, y) whose coordinates are "simultaneous" solutions of a system of inequalities:

$$a_ix + b_iy \leq c \qquad \text{(where } i = 1, 2, \ldots, m\text{).}$$

Each of these inequalities defines a closed half plane. Each of these closed half planes has, in turn, a *boundary line,* defined by the corresponding *equation*:

$$a_ix + b_iy = c_i \qquad \text{(for each } i \text{ above).}$$

In general, a part or even all of a boundary line may belong to the set $\mathscr{K}$. That part of a boundary line which is contained in the set $\mathscr{K}$ is called an *edge*. All the edges taken together form the *boundary* of the polygonal convex set $\mathscr{K}$. These ideas are precisely formulated in the following definition:

DEFINITION 8. (*a*) If $\mathscr{K}$ is a polygonal convex set, the intersection of any boundary line of $\mathscr{K}$ with $\mathscr{K}$ itself is called an edge of $\mathscr{K}$.
(*b*) The union of all edges of $\mathscr{K}$ is called the *boundary* of $\mathscr{K}$.
(*c*) Any point in the boundary of $\mathscr{K}$ is called a *boundary point* of $\mathscr{K}$.

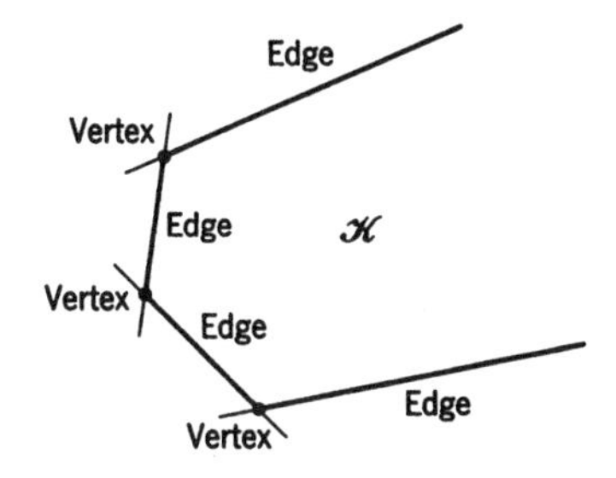

From among the boundary points of a polygonal convex set $\mathscr{K}$, we shall single out those which lie on *more than one* of its boundary lines. We shall call these points the corners or *vertices* of $\mathscr{K}$. They are also often referred to as *extreme points* of $\mathscr{K}$. Certainly, there can be no more of these than there are intersections of pairs of boundary lines and hence there are only a finite number of vertices (possibly none).* We formulate these ideas in the next definition.

DEFINITION 9. An *extreme point* (or *vertex*) of a polygonal convex set $\mathscr{K}$, is a point of $\mathscr{K}$ which is contained in at least two distinct boundary lines.

Because a vertex lies on at least two distinct boundary lines, its coordinates satisfy *at least two* of the equations $a_ix + b_iy = c_i$. Let us say

$$\left.\begin{matrix} a_rx + b_ry = c_r \\ a_sx + b_sy = c_s \end{matrix}\right\} \qquad \text{(for some } r \neq s)\dagger$$

and in addition these coordinates x and y satisfy all the remaining inequalities,

$$a_ix + b_iy \leq c_i \qquad \text{(for } i \neq r, s).$$

* Actually, if $\mathscr{K}$ is the intersection of m closed half planes, we can show that the number of vertices of $\mathscr{K}$ cannot exceed m. In the case $m = 1$, there are no vertices, i.e., a closed half plane has no extreme points.

† These two equalities are, moreover, *not redundant*, because they define distinct lines.

On the other hand, a boundary point which is not a vertex, satisfies exactly *one* of the *equalities* $a_ix + b_iy = c_i$, let us say

$$a_kx + b_ky = c_k,$$

and in addition satisfies all the remaining *strict* inequalities

$$a_ix + b_iy < c_i \qquad (\text{for } i \neq k).^*$$

An edge e of a polygonal convex set $\mathscr{K}$ is the intersection of $\mathscr{K}$ with one of its boundary lines l. As $\mathscr{K}$ and l are both convex sets,

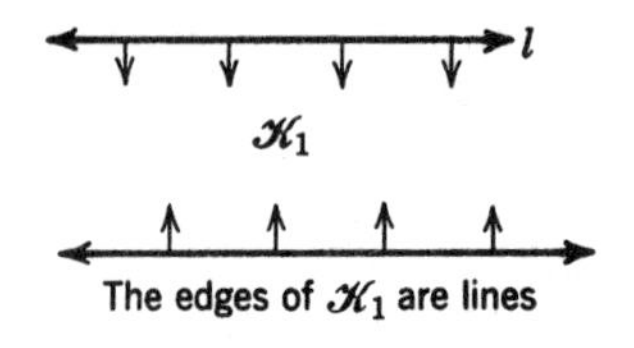

The edges of $\mathscr{K}_1$ are lines

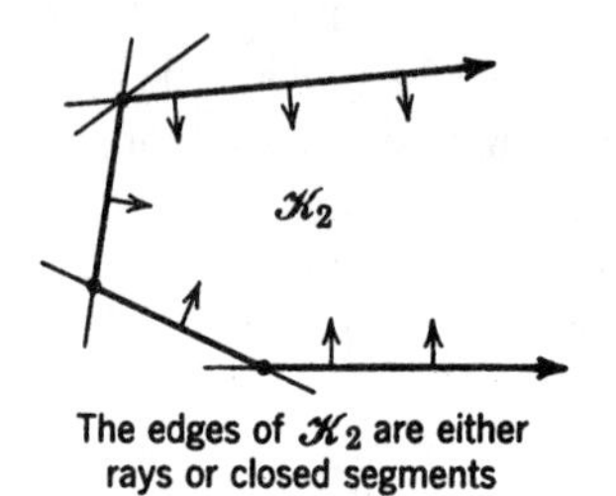

The edges of $\mathscr{K}_2$ are either rays or closed segments

Figure 9

their intersection e must be convex. Therefore, by Theorem 7, e must be one of the following: the line l itself, a ray, a half line, or a segment with or without either endpoint. We shall show below, that e must, in fact, contain its endpoints, if there are any. This reduces the actual possibilities for an edge to

a *line*, a *ray*, or a *closed segment*.

Figure 9 indicates that these configurations actually can occur. We shall also show that an endpoint of an edge is not only contained in that edge, but is also an *extreme point* of $\mathscr{K}$ (vertex of $\mathscr{K}$).

Let the edge e consist of all those points of $\mathscr{K}$ which satisfy

$$a_1x + b_1y = c_1,$$

and let l be the (boundary) line which this equation defines. Let $B = (x_1, y_1)$ be an endpoint of e and let us select the following parametric representation for l:

$$x = x_1 + b_1r \quad \text{and} \quad y = y_1 - a_1r \qquad (\text{for all } r \in R).$$

* This will be true provided the half planes which define $\mathscr{K}$ are all distinct, i.e., provided no two of the equalities $a_ix + b_iy = c$ are redundant.

(See Equation 2, Section 3.) With this particular parametric representation, the endpoint B has the parameter value $r = 0$. Let A be any point of edge e with parameter value a. We assume that $a \neq 0$, for otherwise B is the same as A and the proof will be trivial. We have already shown, in our proof of Theorem 7, that all points between A and B belong to e. Let T be any such point and let t be its associated parameter value. Then the real number t lies between the real numbers a and 0. (More precisely we have either $a < t < 0$ or $0 < t < a$, depending upon whether $a < 0$ or $0 < a$.) If the coordinates of T are $(\bar{x}, \bar{y})$, then clearly

$$\bar{x} = x_1 + b_1 t \quad \text{and} \quad \bar{y} = y_1 - a_1 t.$$

Let us now consider the *remaining* $m - 1$ boundary lines of $\mathscr{K}$ (assuming that there are altogether m of them). Let these be defined by the equalities

$$a_i x + b_i y = c_i \qquad (\text{for } 2 \leq i \leq m).$$

As $T \in K$, we must certainly have

$$a_i \bar{x} + b_i \bar{y} \leq c_i \qquad (\text{for all } i \text{ with } 2 \leq i \leq m),$$

i.e.,

$$a_i(x_1 + b_1 t) + b_i(y_1 - a_1 t) \leq c_i$$

$\therefore$

$$a_i x_1 + b_i y_1 + (a_i b_1 - b_i a_1)t \leq c_i$$

(for all i such that $2 \leq i \leq m$).

But t can be *any* real number in the interval between a and 0, so, within this interval, we may let $t \to 0$ (that is, we may make $|t|$ as small as we please). We thus obtain

$$(1) \qquad a_i x_1 + b_i y_1 \leq c_i \qquad (\text{for all } i \text{ such that } 2 \leq i \leq m),$$

which, along with the fact that equality holds when $i = 1$, signifies that $B = (x_1, y_1) \in \mathscr{K}$, and hence to e, as well, because it is in the intersection of l with $\mathscr{K}$ (see Definition 8).

It remains to show that $B = (x_1, y_1)$ is an *extreme point* of $\mathscr{K}$. We shall accomplish this by showing that at least *one* of the $m - 1$ inequalities (1) is in fact an equality. Suppose, on the contrary, that all of these inequalities are *strict*, i.e., suppose that

$$a_i x_1 + b_i y_1 < c_i \qquad (\text{for all } i \text{ with } 2 \leq i \leq m).$$

Because there are only a finite number $(m - 1)$ of these we can choose a real number $\varepsilon > 0$, so small that the following strict inequalities still hold:

$$\left.\begin{aligned} &a_i x_1 + b_i y_1 + (a_i b_1 - b_i a_1)\varepsilon < c_i \\ \text{and} \quad &a_i x_1 + b_i y_1 - (a_i b_1 - b_i a_1)\varepsilon < c_i \end{aligned}\right\} \quad \text{(for all } i \text{ with } 2 \le i \le m).$$

We rewrite these inequalities as follows:

$$\left.\begin{aligned} &a_i(x_1 + b_1\varepsilon) + b_i(y_1 - a_1\varepsilon) < c_i \\ \text{and} \quad &a_i(x_1 - b_1\varepsilon) + b_i(y_1 + a_1\varepsilon) < c_i \end{aligned}\right\} \quad \text{(for all } i \text{ with } 2 \le i \le m),$$

and we now observe that corresponding to the parameter values $r = \varepsilon$ and $r = -\varepsilon$, there exist two points of l, call them P and Q, respectively, which are also in $\mathscr{K}$ and consequently are both in edge e. But this is impossible, because point $B = (x_1, y_1)$ corresponds to the parameter value $r = 0$ and must, therefore, lie *between* P and Q, contradicting the fact that B is an endpoint of edge e and cannot lie between two points of e. This completes the proof of the assertions we made above. We summarize them as

THEOREM 8. Each edge of a polygonal convex set $\mathscr{K}$ is either a line, a ray, or a closed segment. Its endpoints, if any, are extreme points of $\mathscr{K}$.

We have now gained considerable insight into the nature of the boundary of a polygonal convex set. It is only natural that we should turn next to a study of its "interior." By this we mean the set which remains if we remove all boundary points from the polygonal convex set.

DEFINITION 10. (*a*) An *interior point* of a polygonal convex set $\mathscr{K}$ is any point of $\mathscr{K}$ which is not in the boundary of $\mathscr{K}$.
(*b*) The set of all interior points of $\mathscr{K}$ is called the *interior* of $\mathscr{K}$.

Definition 10 implies that the coordinates (x, y) of an interior point must satisfy all of the *strict* inequalities

$$a_i x + b_i y < c_i \qquad (i = 1, 2, \ldots, m).$$

The interior is, therefore, the intersection of the *open* half planes, rather than the closed half planes, which are used to define the polygonal convex set $\mathscr{K}$.

A polygonal convex set need not have interior points at all. A line, a ray, or a segment (with or without either endpoint), are all examples of such sets.* However, we shall be specially interested in those which do have interior points. We shall call them "regions."

DEFINITION 11. A polygonal convex set which has interior points is called a *polygonal convex region.*

Every point of a polygonal convex region $\mathscr{K}$ is clearly either an interior point or a boundary point. All points of R^2 which are not points of $\mathscr{K}$ will be called "exterior points."

DEFINITION 12. (*a*) An *exterior point* of a polygonal convex set $\mathscr{K}$ is any point of R^2 which is not in $\mathscr{K}$.
(*b*) The set of all exterior points of $\mathscr{K}$ is called the *exterior* of $\mathscr{K}$.

If one wishes to decide whether any particular point (x, y) of R^2 is an interior, exterior, or boundary point of $\mathscr{K}$, one must evaluate each of the expressions $a_i x + b_i y$, which appears as the left member of a defining inequality for $\mathscr{K}$, and then compare this value with the corresponding number c_i, which appears as the right member of that inequality. It, therefore, appears worthwhile to study the way such expressions "behave." We call these expressions "linear forms."

DEFINITION 13. If a and b are real numbers, the expression

$$ax + by$$

is called a *linear form.*

Note that in this definition we do not impose any restriction on the values of a and b, although in most applications a and b are not both zero.

Any linear form can be evaluated at all points of R^2. It defines, therefore, a function f, whose domain is R^2 and whose range consists of real numbers. The value assumed by f at (x, y) will be designated by $f(x, y)$. The reader should have no difficulty proving that the range of f is usually all of R. There is an obvious exception (see Exercise 2 below). The function f is also often called a *linear functional,* the term "functional" referring to any real (scalar)

* We are referring here to a "two dimensional" interior. As "one dimensional" sets, they can be regarded as having "one dimensional" interiors.

valued function whose domain is a "vector space". We shall not use this terminology here.

The values assumed by a linear form are governed by rather stringent laws. The following theorems describe the most important of these laws.

THEOREM 9. LINEARITY PROPERTIES

If $f(x, y)$ is a linear form, then

(a) $$f(mx, my) = mf(x, y), \quad \text{for all real } m.$$

(b) $$f(x_1 + x_2, y_1 + y_2) = f(x_1, y_1) + f(x_2, y_2).$$

Proof

Let $f(x, y) = ax + by$. Then

(a) $$\begin{aligned} f(mx, my) &= a(mx) + b(my) \\ &= m(ax + by) \\ &= mf(x, y). \end{aligned}$$

(b) $$\begin{aligned} f(x_1 + x_2, y_1 + y_2) &= a(x_1 + x_2) + b(y_1 + y_2) \\ &= (ax_1 + by_1) + (ax_2 + by_2) \\ &= f(x_1, y_1) + f(x_2, y_2). \end{aligned}$$

With the aid of these linearity properties, we derive a useful formula for determining the values assumed by a linear form as one proceeds along any given line in R^2.

THEOREM 10. Let $A = (x_1, y_1)$ and $B = (x_2, y_2)$ be distinct points of R^2; let $f(x, y)$ be a linear form; let $f(A) = f(x_1, y_1) = m$ (value of f at A); let $f(B) = f(x_2, y_2) = M$ (value of f at B);

let $$\begin{Bmatrix} x = (1 - t)x_1 + tx_2 \\ y = (1 - t)y_1 + ty_2 \end{Bmatrix} \quad (\text{for } t \in R)$$

be a parametric representation of line $\overleftrightarrow{AB}$. *If* $P = (x, y)$ is any point of line $\overleftrightarrow{AB}$, corresponding to the parameter value t, *then*

$$f(P) = f(x, y) = m + (M - m)t.$$

Proof

$$\begin{aligned} f(x, y) &= f[(1-t)x_1 + tx_2 \,,\ (1-t)\,y_1 + ty_2] \\ &= f[(1-t)x_1 \,,\ (1-t)y_1] + f[tx_2, ty_2] \\ &= (1-t)f(x_1, y_1) + tf(x_2, y_2) \\ &= (1-t)m + tM \\ &= m + (M-m)t. \end{aligned}$$

Q.E.D.

From this theorem we reap a veritable harvest of valuable information about linear forms.

Suppose, first, that $m = M$; then $f(x, y) = m$, *for all* $t \in R$. In that case $f(x, y)$ remains constant all along line $\overleftrightarrow{AB}$:

THEOREM 11. If a linear form assumes the same value at two different points in R^2, it assumes this same value at all points of the line determined by the two given points.

Suppose, next, that $m \neq M$. Now $f(x, y)$ no longer remains constant along line $\overleftrightarrow{AB}$. In fact, if c is any real number whatsoever, we can determine a unique value for t, such that

$$f(x, y) = m + (M - m)t = c.$$

We merely choose

$$t = \frac{c - m}{M - m}.$$

This means that $f(x, y)$ can be made to assume any specified value c, by choosing an appropriate point on line $\overleftrightarrow{AB}$. Moreover this point is uniquely determined by c:

THEOREM 12. If a linear form assumes different values at two points of R^2, then it will assume any specified real value at a suitably chosen point on the line determined by the two given points.

Notice that while Theorem 11 asserts that a nonconstant linear form may not take the same value at two different points of a line, Theorem 12 asserts that it must take any preassigned value at some point. Let us pursue further the manner in which it assumes these values.

Suppose $m < M$. Then $M - m > 0$ and hence f will be a *strictly increasing function of t.* By this we mean the following: let us designate the expression $m + (M - m)t$ by $f(t)$. Then whenever $t_1 < t_2$, $f(t_1) < f(t_2)$. This is easily proved as follows: If $t_1 < t_2$, then $(M - m)t_1 < (M - m)t_2$, because $(M - m) > 0$. Hence $m + (M - m)t_1 < m + (M - m)t_2$, i.e., $f(t_1) < f(t_2)$. In the particular case: $0 < t < 1$, this tells us that

$$f(0) < f(t) < f(1)$$

i.e., $\quad m < f(x, y) < M \quad$ for all points on segment $\overline{AB}$.

In other words, each value assumed by $f(x, y)$ along a segment is intermediate in value, between its values at the ends. Furthermore, of all the values assumed by $f(x, y)$ on the *closed* segment $\overset{\bullet\!-\!\bullet}{AB}$, the minimum value is attained at point A while the maximum is attained at point B, in the case where $m < M$.

In the case where $M < m$, we shall leave it to the reader to establish that f is a *strictly decreasing function* of t, assuming at each point of segment $\overline{AB}$ a unique value intermediate between its values at the ends, the maximum being achieved (in this case) at A, the minimum at B. (See Exercise 4, below.)

The results proved in the last two paragraphs are conveniently summarized as follows:

THEOREM 13. If a linear form assumes different values at two points A and B, then it assumes at each point of the open segment $\overline{AB}$ a unique value intermediate between its values at the endpoints. Its maximum and minimum values on the closed segment $\overset{\bullet\!-\!\bullet}{AB}$ are therefore, assumed at the endpoints.

Exercises

1. Given the polygonal convex set defined by the inequalities

$$\left\{\begin{aligned} x + 3 &\geq 0 \\ y + 5 &\geq 0 \\ x + y + 5 &\geq 0 \\ x + y &\geq 4 \\ x + 18 &\geq 3y \end{aligned}\right\}$$

(*a*) Express each of the inequalities of this system in the form

$$a_i x + b_i y \leq c_i \qquad (i = 1, 2, 3, 4, 5).$$

(*b*) Obtain a parametric representation for each boundary line.
(*c*) Determine the extreme points of the polygonal convex set.
(*d*) Obtain a parametric representation for each edge.
(*e*) Which edges are segments? Which edges are rays?

2. Prove that the range of the linear function f, defined by $f(x, y) = ax + by$, is all of R (the real numbers), whenever a and b are real numbers *not both zero*. What happens when a and b *are* both zero?

3. Given the linear form $3x - 5y$, and segment $\overset{\bullet\!-\!\bullet}{AB}$ joining the points $A = (-2, 1)$ and $B = (4, 2)$. Determine the values assumed by the linear form at A, at B, and at the midpoint of segment $\overset{\bullet\!-\!\bullet}{AB}$. Where on $\overset{\bullet\!-\!\bullet}{AB}$ does the linear form assume its maximum value? Where does it take on its minimum?

4. Complete the proof of Theorem 13, by showing that if a linear function f, has a smaller value at B than at A, then f is a strictly decreasing function of the parameter t.

6. The Fundamental Extreme Point Theorem

Now that we have acquired a clear picture of the behavior of a linear function along any line in R^2, we are in a position to extend this knowledge to more general convex sets. In fact, let $\mathscr{K}$ be a polygonal convex *region* (see Definitions 10 and 11, above). Suppose $\mathscr{K}$ is defined by the m inequalities

$$a_i x + b_i y \leq c_i \qquad (i = 1, 2, \ldots, m).$$

Consider any interior point $A = (x_1, y_1)$ and any line l, which contains this point A. The line l may, or may not, also contain a boundary point. If l *does* contain a boundary point, say $B = (x_2, y_2)$, then there must be at least one value of i, say $i = k$, such that

$$a_k x_2 + b_k y_2 = c_k.$$

But, as A is an interior point, it is also true that

$$a_k x_1 + b_k y_1 < c_k.$$

Therefore, the linear form $f(x, y) = a_k x + b_k y$ assumes at A, a smaller value than c_k (call this value m) while it assumes at B, precisely the value c_k (let us call it M). Hence, by Theorem 13, it assumes at each point of $\overline{AB}$ a smaller value than c_k. This means that *all points of* (the half open) *segment* $\overset{\bullet}{\overline{AB}}$ *are interior points, while B is a boundary point*! Moreover, there also exist on l, points at which $a_k x + b_k y$ assumes values larger than c_k. These are clearly *exterior* points of $\mathscr{K}$. This proves

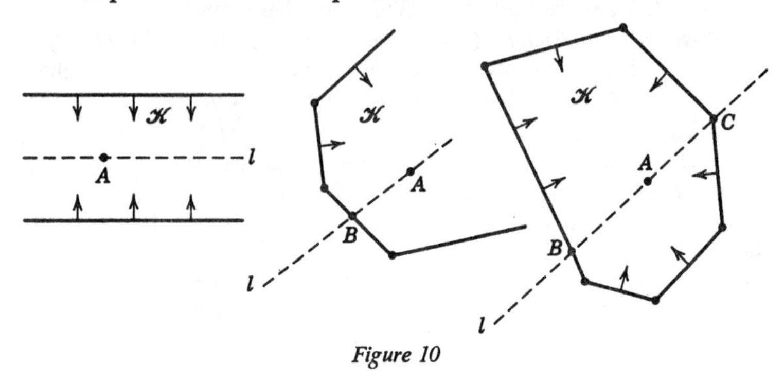

Figure 10

THEOREM 14. If a line l, through an interior point of a polygonal convex region $\mathscr{K}$, intersects the boundary of $\mathscr{K}$, then this line must also contain exterior points.

This very reasonable theorem becomes most illuminating if one forms is *contrapositive*:

THEOREM 15. If a line l does not contain any exterior points of a polygonal convex region $\mathscr{K}$, then *either* it contains only interior points of $\mathscr{K}$ *or*, it contains only boundary points of $\mathscr{K}$.

Thus any polygonal convex region which contains an entire line cannot contain this line "partly inside and partly on the boundary" —the line is either entirely inside or entirely on the boundary as indicated by the first diagram of Figure 10. Such a region is evidently "unbounded," a term we shall presently define.

Let us return now to our discussion just preceding Theorem 14. We saw there that *all points which are between an interior point A and a boundary point B, are themselves interior points.* Now, suppose that line l, through interior point A, intersects the boundary in *more than one* point, say B and C. Then A must lie *between* B and C, for any other possibility would make either B or C an interior point! It now follows that l *cannot intersect the boundary in more than two points.* In fact, the open segment joining any two of these, contains

A and therefore consists entirely of interior points. If there were three points in which l intersects the boundary, one of these three would have to lie between the other two, thereby becoming an interior point. This contradiction excludes the possibility of three or more such points. In summary we have the following:

THEOREM 16. If a line l, through an interior point of a polygonal convex region $\mathscr{K}$, intersects the boundary in more than one point, then the intersection consists of exactly two boundary points. All points between these two boundary points are interior points of $\mathscr{K}$.

In view of Theorem 16, we can classify polygonal convex regions into two main categories:

CATEGORY I. Those polygonal convex regions for which every line through an interior point intersects the boundary in exactly *two* points.

CATEGORY II. Those polygonal convex regions for which there exists at least one line through an interior point intersecting the boundary in *one* point or *no* points.

If $\mathscr{K}$ is a polygonal convex region of Category II, then $\mathscr{K}$ must contain at least one ray (and possibly an entire line) in its interior. It is quite natural to call such a region "unbounded."

On the other hand, a region of Category I, cannot contain a ray within its interior, because Theorem 16 asserts that such a region intersects every line through its interior in a closed segment with endpoints on the boundary. Now, it seems reasonable to suppose that such a region cannot contain a ray in its boundary either. However, we shall not try to prove this supposition, but we shall merely observe that *if such a region $\mathscr{K}$, does not contain a ray in its boundary, then by Theorem 8, its edges are all closed segments,* whose endpoints are the extreme points (vertices) of $\mathscr{K}$. We call such a region $\mathscr{K}$, a "convex polygon." Our precise definition is as follows:

DEFINITION 14. A *convex polygon* is a polygonal convex region $\mathscr{K}$, which has the property that every line containing a point of $\mathscr{K}$, intersects $\mathscr{K}$ in a (closed) segment.

As the edges of a convex polygon are all closed segments, the endpoints of these segments are the vertices of the polygon by Theorem 8. The following theorem is now easily proved:

THEOREM 17. Any convex set which contains the *vertices* of a convex polygon $\mathcal{K}$, contains the *entire* polygon $\mathcal{K}$.

Proof

As the edges of $\mathcal{K}$ are all closed segments, it follows that every boundary point other than a vertex lies on a segment between two vertices. Hence, any convex set S, which contains all the vertices of $\mathcal{K}$, must also contain all the boundary points of $\mathcal{K}$. Moreover, Theorem 16 assures us that every interior point of $\mathcal{K}$ is on (at least one) segment joining two boundary points. Hence, S also contains all interior points of $\mathcal{K}$. This accounts for all points of $\mathcal{K}$, so our proof is complete.

As a special case of this theorem, let us enclose all the vertices of a convex polygon $\mathcal{K}$, in a *rectangle* (see Exercise 4 on page 37). This is always possible, because there are only a finite number of vertices (x_i, y_i), so we can always determine a, b, c, and d, so that $a \leq x_i \leq b$ and $c \leq y_i \leq d$, for all i. It is very natural to call the region $\mathcal{K}$ "bounded," under these circumstances. More generally, we define any bounded set as follows:

DEFINITION 15. A set S (of points in R^2) is called *bounded*, if and only if, there exists a rectangle containing S. (That is, there exist real numbers a, b, c, d, such that, for all $(x, y) \in S$

$$a \leq x \leq b \quad \text{and} \quad c \leq y \leq d.)$$

Any set that is not bounded is called *unbounded*. The regions of Category II are examples of sets which are unbounded. In fact, any set which contains an entire ray is unbounded (see Exercise 3 below). All convex polygons, on the other hand, are bounded.

Our somewhat lengthy analysis of polygonal convex regions has many further ramifications, but we have learned enough to equip us for the task in hand. Our object is to study the values assumed by a linear form $f(x, y) = ax + by$, at the various points of a polygonal convex region $\mathcal{K}$. As a result of this study we shall derive the general theorems which govern the phenomena observed in Part One.

THEOREM 18. THE FUNDAMENTAL EXTREME POINT THEOREM. If $\mathcal{K}$ is a convex polygon and $f(x, y) = ax + by$ is a linear form, then of all the values assumed by $f(x, y)$ at points of $\mathcal{K}$, both the maximum as well as the minimum values, occur at extreme points of $\mathscr{K}$.

Proof

Let $P = (x, y)$ be a point of $\mathscr{K}$, and let us designate the value of f at this point by $f(P)$, i.e., let $f(P) = f(x, y) = ax + by$. Consider any line l, which contains point P. By Definition 14, the line l intersects K in a closed segment. Now, if P is an interior point of $\mathscr{K}$,

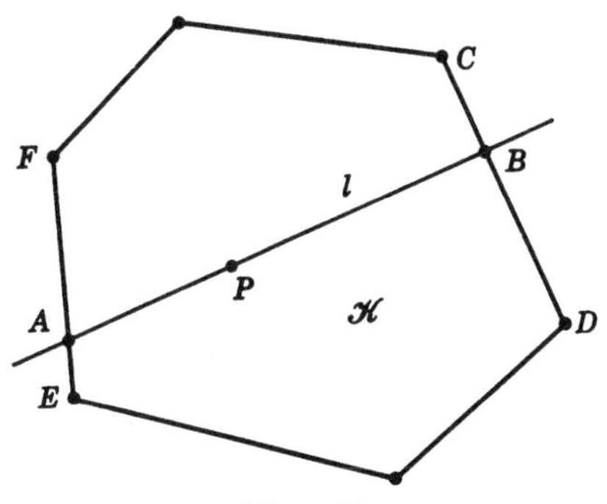

Figure 11

then by Theorem 16, the endpoints A and B of this segment are points of the boundary of $\mathscr{K}$. By Theorem 13, the value $f(P)$ is intermediate between the values $f(A)$ and $f(B)$. Let us say $f(A) \leq f(P) \leq f(B)$. The edges of $\mathscr{K}$ are all closed segments. If B is not an extreme point of $\mathscr{K}$, then by Theorem 8, B must lie between two extreme points C and D and once again, by Theorem 13, the value $f(B)$ is intermediate between the values $f(C)$ and $f(D)$, let us say $f(C) \leq f(B) \leq f(D)$. Therefore, we certainly have $f(P) \leq f(D)$ where D is an extreme point of $\mathscr{K}$. In the same way, if A is not an extreme point of $\mathscr{K}$, then at one end of the edge containing A, say at E, we have $f(E) \leq f(A) \leq f(P)$, i.e., we certainly have

$$f(E) \leq f(P) \leq f(D), \text{ where } D \text{ and } E \text{ are } \textit{vertices} \text{ of } \mathscr{K}.$$

Our argument thus far shows that the value of f at any *interior* point is "matched" by a value at least as great, and a value at least as small, at suitable points of the boundary. The value of f at a *boundary* point is matched, in turn, by a value at least as great and a value at least as small, at suitably chosen *extreme* points. But a convex polygon has a finite number of extreme points. Among the values assumed by $f(x, y)$ at these extreme points, there must consequently be a largest as well as a smallest. These largest and smallest extreme point values are, respectively, the maximum and minimum values attained by $f(x, y)$ on the convex polygon $\mathscr{K}$. This completes the proof of Theorem 18.

Observe that this theorem does not preclude the possibility that f may take the same value at more than one point of $\mathscr{K}$. If this should occur, then by Theorem 11, f must assume the same value at all points of the line determined by these two points, in particular at all points of the closed segment in which this line intersects $\mathscr{K}$. This accounts, for example, for the "multiple optimal solutions" noted in Part One, where a linear form assumed its maximum or minimum value all along an edge of the polygon.

Let us turn our attention next to convex polygonal regions which are not convex polygons. For such a region $\mathscr{K}$ there must be at least one line containing a point of $\mathscr{K}$, which does not intersect $\mathscr{K}$ in a closed segment. But this intersection must be a convex set and it is easy to see that it must therefore be either a line or a ray (see Exercise 4). In either case, we can select a ray, starting say at point P of $\mathscr{K}$ and containing only points of $\mathscr{K}$. If Q is another point on this ray, and if f takes a different value at Q than at P, i.e., if $f(Q) \neq f(P)$, then f will become "arbitrarily large" or "arbitrarily small" along this ray, according as $f(P) < f(Q)$ or $f(Q) < f(P)$, respectively (see Theorem 12). On the other hand, if $f(Q) = f(P)$, then f will retain a constant value all along this ray (see Theorem 11).

Thus, a linear function f need not have a maximum value at all in a polygonal convex region $\mathscr{K}$, if this region is *not* a convex polygon. Similarly, it need not have a minimum value in such region either. However, *whenever it does assume a maximum or a minimum value, it will do so on the boundary of* $\mathscr{K}$. The proof of this is quite similar to the argument used in proving Theorem 17, so we shall merely sketch the idea. We draw a line l through any interior point P. Unless f is constant along this line, it will be a strictly increasing function of the parameter along one of the rays emanating from P, and a strictly decreasing function along the other ray. If neither ray meets the boundary, f will have neither a maximum nor a minimum.

If either ray meets the boundary, then the value of f at this point is at least "as good" (i.e., as large or as small) as the value at P. Hence, if an optimal value (maximum or minimum) exists at all, it will be attained on the boundary. Furthermore, as before, such an optimal value will be "matched" at an extreme point of any edge on which this value is attained. (The only exception will be the case where an edge has no extreme point, i.e., consists of an entire line, and f remains constant all along this line.) The following theorem summarizes these results:

THEOREM 19. If a linear function has either a maximum or a minimum value on a polygonal convex set $\mathscr{K}$, then it assumes such an optimal value on the boundary of $\mathscr{K}$. If this occurs on an edge which has at least one extreme point, then the optimal value is assumed at at least one of these extreme points.

This concludes our discussion of polygonal convex sets in the cartesian plane and the behavior of linear functions on such sets. Specific examples of the various possibilities will be exhibited in several of the exercises below. Some of these possibilities have already been observed in Part One.

Most of the theory developed in this chapter, generalizes readily to any number of variables. This leads to the study of convex sets in an "n-dimensional" space (see Reference 2,3). Regretfully, we must forego this interesting theory in this brief monograph, but fortunately we shall not need it. As we shall see in the next chapter, it is possible to develop a very satisfactory technique for handling linear programming problems involving many variables, without recourse to any of this geometrical theory, interesting though it may be.

Exercises

1. Let $\mathscr{K}$ be defined by

$$\begin{cases} 2x + 3y \leq 6 \\ 2x + 3y \geq -3. \end{cases}$$

(*a*) Prove that this region contains the entire line l, whose parametric representation is

$$x = 1 + 3t \quad \text{and} \quad y = 1 - 2t \qquad (\text{where } t \in R).$$

(*b*) Prove that the linear form $f(x, y) = x + y$ has neither a maximum value, nor a minimum value in $\mathscr{K}$. (*Hint:* express the values assumed by $f(x, y)$ along l, in terms of t.)

(*c*) Prove that the linear form $f(x, y) = 2x + 3y$ remains constant at all points of l.

(*d*) Prove that the linear form $f(x, y) = 4x + 6y$ assumes both a maximum value and a minimum value in the region $\mathscr{K}$. What are these values and where do they occur?

(*e*) Prove that the linear form $f(x, y) = ax + by$ will have both a maximum and a minimum value in $\mathscr{K}$, if and only if, $3a - 2b = 0$. What happens if $3a - 2b \neq 0$?

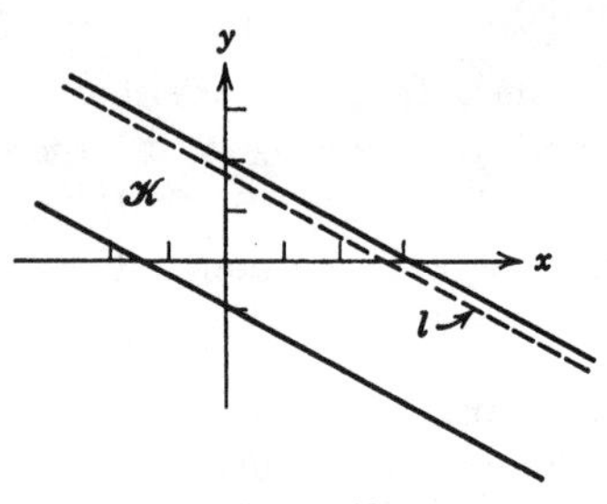

Figure 12

2. Let $\mathscr{K}$ be defined by

$$\begin{cases} -x + y \leq 4 \\ x \geq 0 \\ y \geq 0; \end{cases}$$

also, let l be the line whose parametric representation is:

$$x = 1 + 2t \quad \text{and} \quad y = \tfrac{5}{2} + t \qquad (\text{where } t \in R).$$

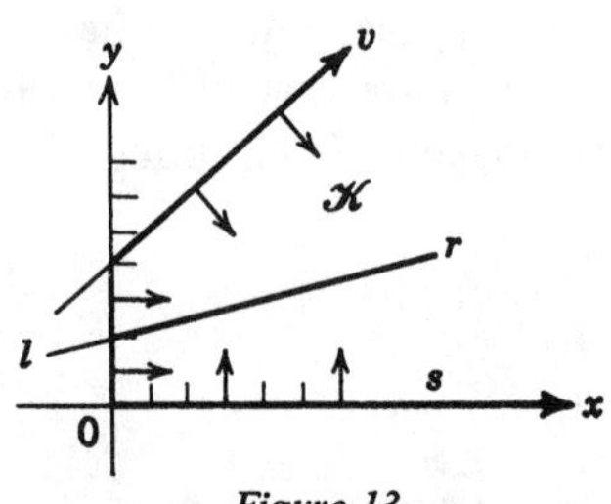

Figure 13

(*a*) Determine where l intersects each of the boundary lines of $\mathscr{K}$.

(*b*) Show that only one of these three intersections is a point of $\mathscr{K}$, and that this particular point corresponds to the parameter value $t = -\frac{1}{2}$.

(*c*) Let the ray r be defined by the parametric representation

$$x = 1 + 2t \quad \text{and} \quad y = \tfrac{5}{2} + t \qquad (\text{where } t \geq -\tfrac{1}{2}).$$

Prove that the entire ray r is contained in $\mathscr{K}$.

(*d*) Show that the linear form $f(x, y) = x - 2y$ retains a constant value at all points of the ray r.

(*e*) Show that the linear form $f(x, y) = x - 2y$ does *not* remain constant along the ray s, defined by the parametric representation

$$x = t \quad \text{and} \quad y = 0 \qquad (\text{where } t \geq 0).$$

In other words, show that $f(x, y)$ varies as one proceeds along the positive x-axis. In fact, show that $f(x, y)$ has a minimum value on this ray, but no maximum value.

(*f*) Show that the linear form $f(x, y) = x - 2y$ has a maximum but no minimum value along the ray v, defined by

$$x = t \quad \text{and} \quad y = 4 + t \qquad (\text{where } t \geq 0).$$

(With the previous result, this shows that $f(x, y)$ has neither a maximum nor a minimum value in $\mathscr{K}$.)

(*g*) Determine whether the linear form $g(x, y) = x + 2y$ has either a maximum or a minimum value in $\mathscr{K}$.

3. Prove that any set which contains an entire ray is unbounded. (*Hint:* Prove that a rectangle cannot contain a ray.)

4. Prove that the intersection of any line l with a polygonal convex set $\mathscr{K}$ is either a line, a ray, or a closed segment. *Hint:* The proof is practically the same as the proof of the first part of Theorem 8 (see pp. 42–43).

***5.** Investigate the question of whether a region of Category I can contain a ray in its boundary. This question may be stated in either of two ways:

(1) If a polygonal convex region does not contain a ray in its interior, then it does not contain a ray in its boundary.

or (contrapositive)

(2) If a polygonal convex region contains a ray in its boundary, then it contains a ray in its interior.

PART THREE

The Simplex Method in Linear Programming

I. Preliminary Remarks

Dantzig's discovery of the Simplex Method ranks high among the achievements of twentieth century applied mathematics. Much of the research in this area makes extensive use of somewhat advanced and highly sophisticated mathematical concepts and techniques. Nevertheless, the Simplex Method for solving linear programming problems can be derived in a strictly elementary fashion using only ideas and techniques familiar to any good high school student or college undergraduate. Underlying the method is the *Gauss-Jordan complete elimination procedure* for solving systems of simultaneous linear equations. This procedure is rapidly gaining popularity because it is essentially a simple, iterative process, readily adapted to computers and easily mastered by even a beginner. Once this procedure is understood, it is a fairly easy transition to the Simplex Method.

2. The Gauss-Jordan Complete Elimination Procedure

We shall introduce the Gauss-Jordan procedure by means of a specific example, but it will be evident that the method is quite general. Consider three equations in five unknowns:

$$
\begin{aligned}
2x - y + 2u - v + 3w &= 14 \\
x + 2y + 3u + v &= 5 \\
x - 2u - 2w &= -10.
\end{aligned}
\tag{1}
$$

Suppose that it is desired to solve these equations for u, v, and w in terms of x and y. We begin by detaching the coefficients, thereby obtaining a representation of the system of equations in "matrix" form:

(2)

x	y	u	v	w	
2	−1	(2)	−1	3	14
1	2	3	1	0	5
1	0	−2	0	−2	−10

Next we eliminate u from each equation except the first. This elimination is initiated by dividing each entry in the first row by 2, the coefficient of u in the first equation. This coefficient is called a "pivot." (It is encircled above.) The effect of this operation is to divide each member of the first equation by the pivot 2. This yields a new first row in which the u-entry is 1:

x	y	u	v	w		
1	$-\frac{1}{2}$	1	$-\frac{1}{2}$	$\frac{3}{2}$	7	(new first row)

(Note that this row represents the equation $x - \frac{1}{2}y + u - \frac{1}{2}v + \frac{3}{2}w = 7$, which is equivalent to the original first equation.) We now multiply each entry in the *new* first row by -3 and add to the corresponding entry of the *previous* second row (in matrix 2). This yields a new second row in which the coefficient of u is zero:

x	y	u	v	w		
1	$-\frac{1}{2}$	1	$-\frac{1}{2}$	$\frac{3}{2}$	7	
-2	$\frac{7}{2}$	0	$\frac{5}{2}$	$-\frac{9}{2}$	-16	(new second row)

Next, we return to the *new* first row, multiply each entry by 2 and add to the previous third row (in matrix 2), thus obtaining a new third row in which the u-entry is again zero. Our matrix now looks like this:

	x	y	u	v	w		
	1	$-\frac{1}{2}$	1	$-\frac{1}{2}$	$\frac{3}{2}$	7	
(3)	-2	$\frac{7}{2}$	0	$\textcircled{\frac{5}{2}}$	$-\frac{9}{2}$	-16	
	3	-1	0	-1	1	4	(new third row)

This new matrix 3, represents a system of linear equations which is completely *equivalent* to the original system (1), i.e., it has the very same solution set. (In fact, in any system of equations, if one replaces an equation by a new one obtained by adding it to a nonzero multiple of any *other* equation of the system, there results an equivalent system.)

The above method of eliminating a variable, such as u, from all equations but one, is called an "*iteration.*" An iteration is inaugurated by selecting a nonzero coefficient of the variable to be eliminated and designating this entry as a *pivot.* (Such a pivot must exist for any variable which is actually present in at least one of the equations.) Dividing its row by the pivot, produces a new row with the entry 1 in the pivotal position. Adding appropriate multiples of this new row to each of the other rows produces new rows with the entry 0 in the pivotal column. This completes the iteration.

Let us now return to our problem. We inaugurate a new iteration, whose purpose this time is to eliminate v from each equation except the second. We select $\frac{5}{2}$ as our new pivot. Dividing each entry of the second row by $\frac{5}{2}$, we obtain a new second row with 1 as the v-entry:

x	y	u	v	w		
$-\frac{4}{5}$	$\frac{7}{5}$	0	1	$-\frac{9}{5}$	$-\frac{32}{5}$	(new second row)

We multiply this new second row by $\frac{1}{2}$ and add to the previous first row (of matrix 3), to get a new first row with 0 as the v-entry:

x	y	u	v	w		
$\frac{3}{5}$	$\frac{1}{5}$	1	0	$\frac{3}{5}$	$\frac{19}{5}$	(new first row)
$-\frac{4}{5}$	$\frac{7}{5}$	0	1	$-\frac{9}{5}$	$-\frac{32}{5}$	

Then we multiply the new second row by 1 and add to the previous third row to obtain a new third row in which the v-entry is again 0. This completes the second iteration and yields the matrix:

	x	y	u	v	w		
	$\frac{3}{5}$	$\frac{1}{5}$	1	0	$\frac{3}{5}$	$\frac{19}{5}$	
(4)	$-\frac{4}{5}$	$\frac{7}{5}$	0	1	$-\frac{9}{5}$	$-\frac{32}{5}$	
	$\frac{11}{5}$	$\frac{2}{5}$	0	0	$-\frac{4}{5}$	$-\frac{12}{5}$	(new third row)

The system of equations represented by this matrix is still equivalent to the original system (1).

A third and final iteration eliminates w from each equation except the third. This is accomplished by dividing the third row of matrix 4 by the qivot $-\frac{4}{5}$, thus making the w-entry unity, and then adding appropriate multiples of this new third row to each of the other rows so as to reduce their w-entries to 0. The final result of this third iteration is the array:

$$\text{(5)} \qquad \begin{array}{ccccc} x & y & u & v & w & \\ \hline \frac{9}{4} & \frac{1}{2} & 1 & 0 & 0 & 2 \\ -\frac{23}{4} & \frac{1}{2} & 0 & 1 & 0 & -1 \\ -\frac{11}{4} & -\frac{1}{2} & 0 & 0 & 1 & 3 \end{array}$$

Our original system of equations (1) is now seen to be equivalent to the following new system:

$$\text{(6)} \qquad \begin{cases} (\frac{9}{4})x + (\frac{1}{2})y + u & = 2 \\ (-\frac{23}{4})x + (\frac{1}{2})y \quad + v & = -1 \\ (-\frac{11}{4})x + (-\frac{1}{2})y \quad + w & = 3. \end{cases}$$

From this we immediately read off our sought for solution, expressing u, v, and w, in terms of x and y:

$$\text{(7)} \qquad \begin{cases} u = 2 - (\frac{9}{4})x - (\frac{1}{2})y \\ v = -1 + (\frac{23}{4})x - (\frac{1}{2})y \\ w = 3 + (\frac{11}{4})x + (\frac{1}{2})y. \end{cases}$$

Clearly, we can generate infinitely many particular solutions to the original system of equations by assigning values to the "parameters" x and y independently. Perhaps the simplest among all these solutions (and most important in the theory of linear programming) is the special solution obtained by letting $x = 0$ and $y = 0$: then $u = 2$, $v = -1$, and $w = 3$. This solution is called a *basic solution* and u, v, and w are called *basic variables* (in the basic solution). The matrix 5 has the advantage of exhibiting explicitly, in its right-hand column, the values of the basic variables, which are readily recognized by the entry 1 in the "unit column vector" directly under each basic variable.

The solution we have just obtained is not the only basic solution. Other basic solutions can be obtained by assigning the value zero to any two of the five variables (these two become "nonbasic") and then determining corresponding values for the remaining three variables (these are now "basic"). All this is conveniently accomplished by utilizing the Gauss-Jordan procedure to solve for the three basic variables in terms of the remaining two nonbasic variables.

In general, if we have m equations in $m + n$ variables, we can designate n of these variables as nonbasic and solve for the remaining basic variables. In most actual applications there will be a unique solution for the basic variables in terms of the nonbasic ones. The Gauss-Jordan procedure then leads to a matrix such as (5) above, wherein the entries under each basic variable constitute a "basic unit column vector." This is simply a column with exactly one entry equal to 1 and the others equal to 0. Each basic variable has the nonzero entry (i.e., the 1) in a different row from all the others. The value of this basic variable, in the associated basic solution, appears in this row in the far right column.

3. The Extended Simplex Tableau

We proceed now to apply these ideas to the solution of linear programming problems. Let us begin by examining a particular one:

Ronald's mathematics teacher has given his class three very long lists of problems with the instruction to submit no more than 100 of them (correctly solved) for credit. The problems in the first set are worth 5 points each, those in the second set are worth 4 points each, and those in the third set are worth 6 points each.

Ronald knows from experience that he requires on the average 3 minutes to solve a 5-point problem, 2 minutes to solve a 4-point problem, and 4 minutes to solve a 6-point one. Because he has other subjects to worry about, he cannot afford to devote more than $3\frac{1}{2}$ hours altogether to this mathematics assignment. Moreover, the first two sets of problems involve numerical calculations, and he knows that he cannot stand more than $2\frac{1}{2}$ hours of work on this type of problem.

Under these circumstances, how many problems in each of the three categories shall he do in order to get the maximum possible credit for his efforts?

Analysis and Solution

Let x = the number of 5-point problems to be done,
y = the number of 4-point problems to be done,
z = the number of 6-point problems to be done.

Ronald's dilemma may now be formulated as follows:

Determine: $x \geq 0, \quad y \geq 0, \quad z \geq 0$

so that:
$$\begin{cases} x + y + z \leq 100 \\ 3x + 2y + 4z \leq 210 \\ 3x + 2y \leq 150 \end{cases}$$

and so that: $M = 5x + 4y + 6z$ is a MAXIMUM.

By introducing additional "slack" variables, u, v, w, we can reformulate the problem as follows:

Determine: $x \geq 0, \quad y \geq 0, \quad z \geq 0, \quad u \geq 0, \quad v \geq 0, \quad w \geq 0$

so that:
$$\begin{cases} x + y + z + u = 100 \\ 3x + 2y + 4z + v = 210 \\ 3x + 2y + w = 150 \end{cases}$$

and so that: $M = 5x + 4y + 6z + 0u + 0v + 0w$ is a MAXIMUM.

We shall find it helpful to treat the last equation similar to the other equations, so we transpose all terms of this equation to the left side and express our problem in this manner:

Among all the possible solutions of the system of equations

$$(1) \qquad \begin{cases} x + y + z + u = 150 \\ 3x + 2y + 4z + v = 210 \\ 3x + 2y + w = 150 \\ -5x - 4y - 6z + M = 0 \end{cases}$$

determine a solution for which

$$(2) \qquad x \geq 0, \quad y \geq 0, \quad z \geq 0, \quad u \geq 0, \quad v \geq 0, \quad w \geq 0$$

and for which

(3) M is as large as possible!

The following bit of terminology has become fairly standard in linear programming and will be found quite helpful in our discussion:

DEFINITION. Any solution of the system of equations (1) which also satisfies (2) is called a *feasible solution.*

DEFINITION. Any solution of the system of equations (1) which also satisfies (3) is called an *optimal solution.*

The problem is to determine a feasible solution which is also optimal! A fundamental existence theorem states that whenever there exists an optimal, feasible solution, there exists one which is also basic.* The Simplex Procedure is a method for arriving systematically at such an optimal, basic feasible solution, when it exists, or revealing that there is none, when one does not exist.

The method begins by writing the array (called a *Simplex Tableau*) corresponding to the system (1):

x	y	z	u	v	w	M	
1	1	1	1	0	0	0	100
3	2	4	0	1	0	0	210
3	2	0	0	0	1	0	150
−5	−4	−6	0	0	0	1	0

The presence of basic unit column vectors in the last four columns under the variables, signifies that we already have a basic feasible solution to start with, namely

$$x = 0, y = 0, z = 0, u = 100, v = 210, w = 150, \quad \text{and} \quad M = 0.$$

However, this solution is not optimal because (reading from the last row) we have the equation

$$M = 0 + 5x + 4y + 6z - 0u - 0v - 0w,$$

and clearly we can increase the value of M by increasing either x, y, or z. Now at the moment each of these three (nonbasic) variables has the value 0. We are free to choose any of the three and allow it to assume a positive value. It appears advantageous to select z for this purpose, because each unit increase in the value of z increases the value of M by 6 units, whereas a corresponding unit increase in either x or y, increases M by only 4 or 5 units, respectively.

So we allow z to increase, *keeping x and y equal to* 0. We observe at once that this change in the value of z, not only increases the value of M, but forces us to reduce the values of u and v, two of the variables which were temporarily in the basis. This reduction must occur because u, v, and w must satisfy the three equations represented

* See Reference 4, Chapter 4, or Reference 8, Chapter 3.

by the first three rows of the matrix, i.e.:

$$\begin{aligned} u &= 100 - x - y - z \\ v &= 210 - 3x - 2y - 4z \\ w &= 150 - 3x - 2y, \end{aligned}$$

and with x and y both equal to zero these equations become

$$\begin{aligned} u &= 100 - 1z \\ v &= 210 - 4z \\ w &= 150 - 0z. \end{aligned}$$

Clearly, any increase in z reduces u and v (but does not affect w in this particular example). Now remember that u, v, and w are also constrained to remain non-negative. Hence, we cannot increase z indefinitely. In fact, the first equation permits us to increase z by no more than $100/1 = 100$ units, while the second equation permits us to increase z by no more than $210/4 = 52\frac{1}{2}$ units. (The third equation in this case permits an unlimited increase in z.) Clearly then, the allowable increase in z may not exceed the *smaller* of the two ratios 100/1 and 210/4, namely $52\frac{1}{2}$. This change in the value of z (from 0 to $52\frac{1}{2}$) reduces the value of u from 100 to $47\frac{1}{2}$, reduces the value of v from 210 to 0, and leaves the value of w unchanged at 150. We now have a *new* basic feasible solution

$$x = 0, y = 0, v = 0, u = 47\tfrac{1}{2}, z = 52\tfrac{1}{2}, w = 150 \quad \text{and} \quad M = 315.$$

In this new basic feasible solution, the value of M is substantially greater than it was before.

Now it is not at all necessary to use this cumbersome line of reasoning in order to obtain the new (improved) basic feasible solution. We have already seen in the preceding section that one can arrive at a new B.F.S. by systematically transforming the matrix with the aid of the Gauss-Jordan complete elimination procedure. Thus in the present illustration, we do this using the entry 4 as a pivot. This pivot is selected on the basis of the following considerations: locate the "most negative" entry in the last row;* this locates the variable to be introduced into the basis; divide each *positive* element in the column under this variable (and only positive elements) into the extreme right member of its row; select as a

* Actually, any negative entry in the last row may be used, but this one yields the largest increase in M per unit increase in the variable associated with this entry. It is not always the best choice.

pivot the divisor which yields the smallest quotient; use this pivot to perform a complete iteration. In the present example, this yields the following new matrix:

x	y	z	u	v	w	M		
$\frac{1}{4}$	$\frac{1}{2}$	0	1	$-\frac{1}{4}$	0	0	$47\frac{1}{2}$	$= u$
$\frac{3}{4}$	$\frac{1}{2}$	1	0	$\frac{1}{4}$	0	0	$52\frac{1}{2}$	$= z$
3	2	0	0	0	1	0	150	$= w$
$-\frac{1}{2}$	-1	0	0	$\frac{3}{2}$	0	1	315	$= M$

From this new matrix we read off at a glance the new (improved) B.F.S. The four unit vectors under the headings z, u, w, and M, identify these as the basic variables whose new values are displayed in the column at the right.* The nonbasic variables, at this point, are x, y, and v.

Although this B.F.S. is an improvement over the previous one, the presence of negative numbers in the last row signifies that it is possible to increase M still further. The "most negative" entry is -1. We note that all the entries in this column are positive, so we calculate all three ratios

$$\frac{47\frac{1}{2}}{\frac{1}{2}} = 95, \qquad \frac{52\frac{1}{2}}{\frac{1}{2}} = 105, \qquad \frac{150}{2} = 75.$$

The *smallest* ratio, 75, identifies 2 as the next pivot. Using this pivot we perform another iteration and obtain

x	y	z	u	v	w	M		
$-\frac{1}{2}$	0	0	1	$-\frac{1}{4}$	$-\frac{1}{4}$	0	10	$= u$
0	0	1	0	$\frac{1}{4}$	$-\frac{1}{4}$	0	15	$= z$
$\frac{3}{2}$	1	0	0	0	$\frac{1}{2}$	0	75	$= y$
1	0	0	0	$\frac{3}{4}$	$\frac{1}{2}$	1	390	$= M$

This matrix yields a new B.F.S. with an increased value of M. Moreover, *no further improvement* in the value of M is possible,

* This column, together with the identification of the basic variables next to it, is called the "stub." It is often placed on the left side of the matrix, instead of at the extreme right as we are doing here.

since there are no longer any negative entries in the last row. We have at last attained our desired *optimal solution,*

$$x = 0,\ y = 75,\ z = 15,\ u = 10,\ v = 0,\ w = 0, \quad \text{and} \quad M = 390.$$

Interpreting this solution, we see that Ronald's best bet is to do *none* of the problems in the first set, 75 problems in the second set (at only 4 points each), and 15 problems in the third set (at 6 points each). This secures for him a total of 390 points, the maximum credit possible under the conditions set forth in the problem. Observe that some "slack" remains in the optimal solution, for he actually does only 90 problems (not 100). This shows up in the fact that the slack variable u remains in the final basis with a value of 10. However, Ronald does have to use all of his available time, namely 210 minutes, and he is forced to put up with the full 150 minutes on numerical calculation type problems, which he dislikes.

In practice, one usually omits the column headed M, as this column never changes throughout the computation. With this slight alteration, the following summarizes our solution of the above problem by the Simplex Method:

	x	y	z	u	v	w			Comments
$\frac{100}{1} = 100$	1	1	1	1	0	0	100	$= u$	*Initial B.F.S.*
$\rightarrow \frac{210}{4} = 52\frac{1}{2}$	3	2	(4)	0	1	0	210	$= v$	Choose $x = y = z = 0$; u, v, w are in the basis and have the values indicated. The most negative entry in the last row is -6 and the entry 4 (in the circle) becomes the pivot.
	3	2	0	0	0	1	150	$= w$	
	-5	-4	-6 ↑	0	0	0	0	$= M_0$	

	x	y	z	u	v	w			Comments
$\frac{47\frac{1}{2}}{\frac{1}{2}} = 95$	$\frac{1}{4}$	$\frac{1}{2}$	0	1	$-\frac{1}{4}$	0	$47\frac{1}{2}$	$= u$	*New B.F.S.*
$\frac{52\frac{1}{2}}{\frac{1}{2}} = 105$	$\frac{3}{4}$	$\frac{1}{2}$	1	0	$\frac{1}{4}$	0	$52\frac{1}{2}$	$= z$	z enters the basis displacing v. There are still negative entries in the last row, indicating that M' is not maximum. The new pivot is 2.
$\rightarrow \frac{150}{2} = 75$	3	(2)	0	0	0	1	150	$= w$	
	$-\frac{1}{2}$	-1 ↑	0	0	$\frac{3}{2}$	0	315	$= M'$	

	x	y	z	u	v	w			Comments
	$-\frac{1}{2}$	0	0	1	$-\frac{1}{4}$	$-\frac{1}{4}$	10	$= u$	*New B.F.S.*
	0	0	1	0	$\frac{1}{4}$	$-\frac{1}{4}$	15	$= z$	y enters the basis displacing w. There are no more negative entries in the last row. This indicates that M'' is *now a maximum.*
	$\frac{3}{2}$	1	0	0	0	$\frac{1}{2}$	75	$= y$	
	1	0	0	0	$\frac{3}{2}$	$\frac{1}{2}$	390	$= M''$	

Notice that the selection of a pivot for each new iteration requires examining the entries directly above some negative last-row entry and dividing only those that are *positive* into the corresponding entries at the extreme right. The reason we ignore those that are negative or zero, is worth repeating. If a variable has a negative or a zero coefficient in any particular equation, then that variable may be assigned as large a value as one pleases, and it will still be feasible as far as that equation is concerned. Now if this should occur in every equation (i.e., if all the entries above *some* negative last-row entry are negative or zero) then M can be made as large as one chooses by simply taking arbitrarily large values for the variable associated with these entries. In this case we say that M has an "infinite maximum."

Many linear programming problems call for *minimizing* rather than maximizing a linear expression similar to M above. This is easily accomplished by maximizing $(-M)$ using the Simplex Method we have already developed. For example, in the above problem $(-M)$ represents the expression $-5x - 4y - 6z$. If we write this as the equation

$$5x + 4y + 6z + (-M) = 0$$

and use this new variable $(-M)$ in place of the variable M, the initial matrix looks like this:

x	y	z	u	v	w	$(-M)$		
1	1	1	1	0	0	0	100	$= u$
3	2	4	0	1	0	0	200	$= v$
3	2	0	0	0	1	0	150	$= w$
5	4	6	0	0	0	1	0	$= (-M)$

This initial basic feasible solution is *already optimal*, because there are no negative entries in the last row. The maximum value of $(-M)$ is zero and hence the *minimum* value of M is also zero. This result was, of course, obvious from the start, but it is worth seeing that it is obtainable by our Simplex Method. We shall consider a less trivial illustration in the next section.

4. The Use of Artificial Variables

In the previous section an initial basic feasible solution was available right from the beginning due to the presence, within the initial matrix, of a set of basic unit column vectors, one for each basic variable. These are so handy, that it pays to introduce them "artificially" when they are not already present.

We shall describe how this can be done by means of a rather simple problem which, we admit, can be easily solved by other methods, but which we shall use as a vehicle to introduce the artificial variable technique.

Mr. Hy. P. Kondriak has been ordered by his physician to take daily at least 24 units of vitamin B_1 and at least 25 units of vitamin B_2. Unfortunately, these are not available in pure form, but the local drug stores sell HEALTH tablets at one cent each and STRENGTH capsules at three cents apiece. Each tablet contains 1 unit of B_1 and 5 units of B_2, while each capsule contains 4 units of B_1 and 1 unit of B_2.

Under these conditions, how many HEALTH tablets and how many STRENGTH capsules should Mr. Hy. P. Kondriak purchase daily in order to obtain the required vitamins at *minimum cost*?

Analysis and Solution

Let x = number of HEALTH tablets to be purchased,
y = number of STRENGTH capsules to be purchased.

Mr. Hy. P. Kondriak's problem may be formulated as follows:

Determine: $x \geq 0, \quad y \geq 0$

so that:
$$\begin{cases} x + 4y \geq 24 \\ 5x + y \geq 25 \end{cases}$$

and so that: $m = x + 3y$ is a MINIMUM.

By introducing "slack" variables u and v (to change the inequalities to equations) and by letting $M = (-m)$, we can reformulate the problem as follows:

Determine: $x \geq 0, \quad y \geq 0, \quad u \geq 0, \quad v \geq 0$

$$\text{so that:} \qquad \begin{cases} x + 4y - u \qquad = 24 \\ 5x + \ y \qquad - v = 25 \end{cases}$$

and so that: $M = -x - 3y + 0u + 0v$ is a MAXIMUM.

Now unfortunately, the negative coefficients preceding u and v prevent these variables from supplying the very desirable unit basis vectors when setting up the initial matrix. So we introduce two new "artificial" variables a and b, into the equations, as follows:

$$\begin{cases} x + 4y - u \qquad + a \qquad = 24 \\ 5x + \ y \qquad - v \qquad + b = 25. \end{cases}$$

Ordinarily, this would change drastically the character of our problem, but we prevent this by the following ingenious device: we subtract from M the quantity $pa + pb$, *where* p *is an unspecified but "large" positive number*; call this result M', i.e.,

$$M' = M - pa - pb = -x - 3y - pa - pb,$$

and rewrite this equality as follows:

$$x + 3y + pa + pb + M' = 0.$$

We then proceed to seek a maximum value for M' instead of M. In other words, we are going to try to solve the following new "*artificial*" problem:

Determine: $x \geq 0, \quad y \geq 0, \quad u \geq 0, \quad v \geq 0, \quad a \geq 0, \quad b \geq 0$

$$\text{so that:} \qquad \begin{cases} x + 4y - u \qquad + a \qquad\qquad = 24 \\ 5x + \ y \qquad - y \qquad + b \qquad = 25 \\ x + 3y \qquad\qquad + pa + pb + M' = \ 0 \end{cases}$$

and so that: M' is as large as possible.

If a basic feasible solution exists for our original problem, then it will constitute a basic feasible solution for our artificial problem, with $a = b = 0$. Therefore, if we apply the Simplex Method to the artificial problem, we must eventually arrive at such a basic feasible solution (with $a = b = 0$) because the large negative coefficient $-p$ prevents M' from reaching its maximum so long as either $a > 0$ or $b > 0$. (If it should happen that M' did reach a maximum before a and b were both eliminated from the basis, this could only mean that the original problem was not feasible.) Having attained to a B.F.S. with $a = b = 0$, we can drop the artificial variables altogether, and continue with the original variables, applying the Simplex Method until either M attains a maximum value, or we ascertain that M has an infinite maximum.

The starting matrix for our artificial problem looks like this:

x	y	u	v	a	b	M'	
1	4	−1	0	1	0	0	24
5	1	0	−1	0	1	0	25
1	3	0	0	0	0	1	0
				p	p		

(There is no B.F.S. evident at this stage.)

For convenience in subsequent calculation, each entry in the last row has been separated into two parts, the part involving p being written below the part which does not involve p. In each column of this "double row" the two parts form a single entry, and the double row still represents a single equation of the artificial problem.

As we have indicated, this matrix does not yield an obvious B.F.S., but we can readily obtain one that does. We simply multiply *each* of the first two rows by $-p$ and add both of these results to the last row. It is convenient to record these totals in the extra space which has been provided in the last row, because they involve p. We now have the following equivalent representation for our artificial problem:

	x	y	u	v	a	b	M'		
$\frac{24}{1} = 24$	1	4	−1	0	1	0	0	24	$= a$
→ $\frac{25}{5} = 5$	⑤	1	0	−1	0	1	0	25	$= b$
	1	3	0	0	0	0	1	0	$= M'$
	$-6p$ ↑	$-5p$	$1p$	$1p$				$-49p$	

(Initial B.F.S.) Both artificial variables are in the basis.

The appearance of three basic unit column vectors under a, b, and M', heralds the presence of our first B.F.S. for the artificial problem.

We inaugurate a new iteration in the usual manner, using the "most negative" last row entry ($-6p$) to locate the new pivot (5), and x now enters the basis, displacing the artificial variable b. Notice that the two parts of the double row can be transformed as if they were separate rows. Notice also, that we do not bother to transform the columns headed a, b, and M', because the latter column never changes anyway, and because the artificial variables never reenter the basis once they are displaced.

	x	y	u	v	a	b	M'			
$\rightarrow \frac{19}{\frac{19}{5}} = 5$	0	$\textcircled{\frac{19}{5}}$	-1	$\frac{1}{5}$				19	$= a$	
$\frac{5}{\frac{1}{5}} = 25$	1	$\frac{1}{5}$	0	$-\frac{1}{5}$				5	$= x$	*New B.F.S.* The artificial variable "a" is still in the basis.
	0	$\frac{14}{5}$	0	$\frac{1}{5}$				-5	$= M'$	
	0	$-\frac{19}{5}p$	$1p$	$-\frac{1}{5}p$				$-19p$		
		$\uparrow$								

The large negative entry $-\frac{19}{5}p$ signifies that further improvement in M' is still possible. We perform another iteration using the pivot $\frac{19}{5}$:

x	y	u	v	a	b	M'			
0	1	$-\frac{5}{19}$	$\frac{1}{19}$				5	$= y$	
1	0	$\frac{1}{19}$	$-\frac{4}{19}$				4	$= x$	*New B.F.S.* No artificial variables in this basis.
0	0	$\frac{14}{19}$	$\frac{1}{19}$				-19	$= M'$	
0	0	0	0				0		

This time y has entered the basis displacing the other artificial variable a. The number p no longer influences the value of M'. We have arrived at a basic feasible solution to our *original* problem

$$x = 4, y = 5, u = 0, v = 0, m = 19.$$

(Note that the value of m is now simply the negative of the value of M', because the artificial variables a and b have vanished.)

This B.F.S. also happens to be optimal in this particular problem, as there are no further negative entries in the M' row (the double row). In more complicated problems, this need not be so. After completing the "first phase" in which the artificial variables are elimininated, further iterations may still be in order if there are negative entries remaining in the final row. Then one continues to apply the Simplex procedure until either M is maximized, or one ascertains that it has an infinite maximum. In the former case m is minimized. In the latter case m has an "infinite minimum." In this "second phase" one is actually working with basic feasible solutions to the *original* problem, the artificial problem having served its purpose during the "first phase." This purpose was to lead us to an *initial* B.F.S. for the *original* problem.

5. The Condensed Simplex Tableau

A further streamlining of the Simplex Method was devised in the mid 1950's by S. Vajda and M. Beale, and has been used extensively by A. W. Tucker.* This method, called the Condensed Tableau, eliminates from the regular Simplex Tableau as many columns as there are basic variables.

To see how this is accomplished, let us look back at the sequence of tableaux on page 68. The first iteration removes from the basis the variable v, and replaces it by the variable z. In the process, the column of entries under z is transformed into a *basic unit column vector* $\begin{bmatrix} 0 \\ 1 \\ 0 \\ 0 \end{bmatrix}$, while a similar basic unit column vector under v is replaced by a new set of entries. The Condensed Tableau Method eliminates entirely all such basic unit vector columns (along with their basic variable headings). It transforms the column of z-entries directly into the new column of v-entries and merely interchanges the labels z and v. Two simple rules determine these new entries as well as the remaining entries of the transformed tableau. We shall derive these rules and then state them. Although we shall refer to the variables v and z which are interchanged by the first iteration, it will be apparent that the discussion is quite general and applies to each iteration.

Prior to the iteration, the pivot p $(= 4$ in the particular illustration we are using) appears in the z-column and is located in the same row

* "Condensed Schemata for Dantzig's Simplex Method," Memorandum by A. W. Tucker, for SOCIALØKONOMISK INSTITUTT, University of Oslo, 1959.

as the entry 1 of the v-column. The current value of z is zero, and the current value of the basic variable v (namely 210) appears at the extreme right in this row. We indicate all the items of interest, schematically as follows:

	z		any other column		v			
	⋮		⋮		⋮			
pivotal row	… (p)	…	q	…	1	…		= v (variable in basis —to be replaced by z)
	⋮		⋮		⋮			
any other row	… r	…	s	…	0	…		
	⋮		⋮		⋮			

The iteration begins with division of the pivotal row by p. The new value at the extreme right of this row is the new value of z.

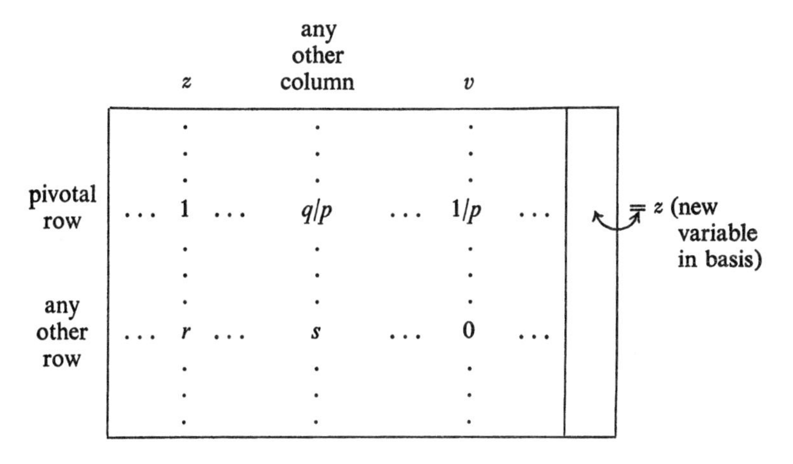

Now consider any other row with entry r in the pivotal column (i.e., in the z-column). To *each* entry of this row is added the product of $(-r)$ by the *corresponding* entry of the pivotal row. The result is the new matrix:

		z		any other column		v		
pivotal row	$\ldots$	1	$\ldots$	q/p	$\ldots$	$1/p$	$\ldots$	$= z$
any other row	$\ldots$	0	$\ldots$	$s - r(q/p)$	$\ldots$	$-r/p$	$\ldots$	

In this matrix, the z-column is now a basic unit vector. We drop this column entirely and move the v-column into its place. We also rewrite $s - r(q/p)$ in the form $s + q(-r/p)$. Our tableau now takes on the following appearance:

(Condensed Tableau *After* Iteration)

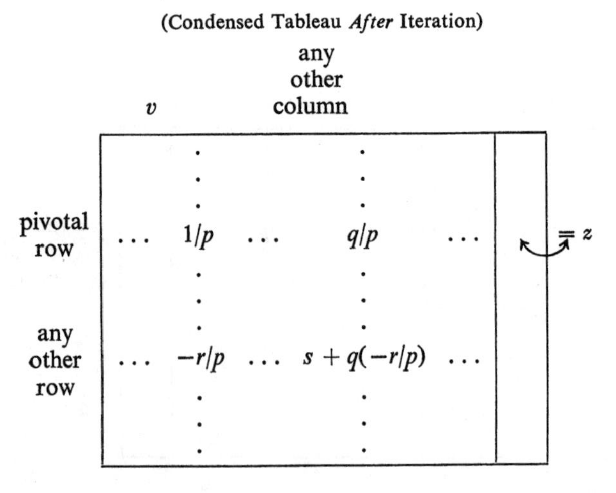

Finally, let us condense the original tableau (before the iteration) by dropping the (basic unit vector) v-column:

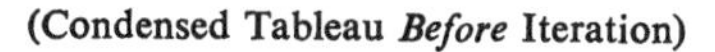

(Condensed Tableau *Before* Iteration)

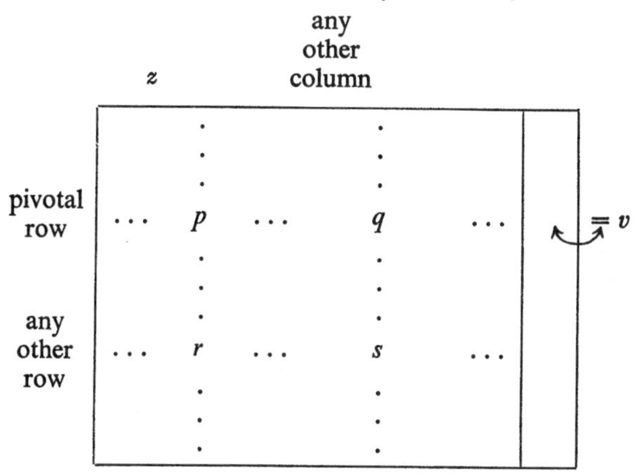

Comparing these two condensed tableaux we now obtain the following rules for transforming the original matrix into the new one:

RULE 1. In the horizontal row of the pivot, replace the pivot by its reciprocal and divide all the remaining entries of this row by the pivot. In the vertical column of the pivot divide all entries other than the pivot by the negative of the pivot.

RULE 2. Add to each remaining entry in the old tableau, the product of the old entry in the same column (to the right or left of the pivot) and the new entry in the same row (above or below the pivot).

As an example we exhibit the solution of the aforementioned problem by the Condensed Tableau Method. Compare this with the Extended Simplex Tableaux on pages 65, 67, 68.

	x	y	z		
$100/1 = 100$	1	1	1	100	$= u$
$210/4 = 52\frac{1}{2} \rightarrow$	3	2	(4)	210	$= v$
	3	2	0	150	$= w$
	-5	-4	-6 ↑	0	$= M$

(Exchange z and v on the first iteration)

	x	y	v			
$47\frac{1}{2}/\frac{1}{2} = 95$	$\frac{1}{4}$	$\frac{1}{2}$	$-\frac{1}{4}$	$47\frac{1}{2}$	$= u$	(Exchange y and w on the second iteration)
$52\frac{1}{2}/\frac{1}{2} = 105$	$\frac{3}{4}$	$\frac{1}{2}$	$\frac{1}{4}$	$52\frac{1}{2}$	$= z$	
$150/2 = 75 \rightarrow$	3	(2)	0	150	$= w$	
	$-\frac{1}{2}$	-1	$\frac{3}{2}$	315	$= M$	
		$\uparrow$				

x	w	v			
$-\frac{1}{2}$	$-\frac{1}{4}$	$-\frac{1}{4}$	10	$= u$	
0	$-\frac{1}{4}$	$\frac{1}{4}$	15	$= z$	Final (Optimal) Tableau
$\frac{3}{2}$	$\frac{1}{2}$	0	75	$= y$	
1	$\frac{1}{2}$	$\frac{3}{2}$	390	$= M$	

The reader should check each entry using the appropriate rule given above. He will do well to solve the problems presented in the exercises at the end of this chapter using both the Extended Simplex Tableau as well as the Condensed Tableau Method. In this way he will gain both skill and insight into the methods of linear programming.

We shall return to these techniques after we have discussed the Theory of Games.

6. Degeneracy

In the Simplex Method we proceed systematically from one B.F.S. to another. Each new B.F.S. is obtained by selecting a variable not in the basis, to replace one that is. The criterion for making the selection consists of examining the entries directly above some negative last-row entry and dividing *only those that are positive* into the corresponding entries at the extreme right of the matrix. The latter entries represent values of variables, currently in the basis. *The minimum of the ratios so obtained, represents the value of the new variable after it is introduced into the basis.* This value appears in the right hand column at the end of the iteration, along with the new values assumed by the variables remaining in the basis (see Section 3 above). Two possible difficulties can arise here:

(*a*) *If the minimum ratio happens to be zero* (*this can happen only if one of the variables currently in the basis has the value* 0), then the "new" variable does not change its value at all, as it enters the basis, since its original value as a nonbasic variable was 0, and its "new" value is also 0. But then the variables remaining in the basis do not change their values either! In short we really do not have a *new* basic feasible solution, but merely a repetition of the old one, except that one of the basic variables with a zero value has been named "nonbasic," and in its place one of the nonbasic variables (which of course is also zero) has been named "basic" in its place. Clearly, the iteration produces no improvement in the value of the objective function M, when this happens.

(*b*) *If there is a "tie" between two* (*or more*) *of the above mentioned ratios for a minimum value,* then when the new variable enters the basis and takes on this value, it reduces to zero not only the variable which it is displacing, but also any other variable remaining in the basis whose original value (right hand entry) was used in computing one of these minimum ratios.* In both cases (*a*) and (*b*) *one* (*or more*) *of the variables in the basis has the value zero.* This phenomenon is known as *degeneracy*. A B.F.S. in which some of the basic variables

* This is readily proved algebraically. Suppose the entries under discussion appear in the matrix as follows:

$$\begin{array}{c|c} \ldots a \ldots\ldots & R \\ \ldots\ldots\ldots\ldots & \ldots \\ \ldots a' \ldots\ldots & R' \end{array} \quad \left(\text{where: } \frac{R}{a} = \frac{R'}{a'}\right).$$

If we select, let us say, a, as the pivot, then dividing its row by a, yields a new row with the entries

$$\ldots 1 \ldots\ldots \Big| \ \frac{R}{a} \ \text{(value of new Variable entering the basis).}$$

In the iteration, we then multiply this row by $-a'$ and add it to the other row thus obtaining

$$\ldots 0 \ldots\ldots \Big| \ R' - \frac{a'R}{a} = 0 \ \text{(new value of variable remaining in the basis).}$$

are zero (in addition to the nonbasic variables, which are zero by definition) is called a *degenerate basic feasible solution.*

When degeneracy occurs, then, as demonstrated in case (*a*), the next iteration may produce no improvement in M. This is, of course, disconcerting but seldom fatal in practice. Actually, degeneracy is handled in most cases by simply ignoring it and continuing to introduce new variables into the basis. The value of M can remain stationary through several such iterations and then begin to increase again.

In rare cases it is possible to return to a basis which has already been encountered in a previous iteration. This complication is known as "cycling." Various methods have been devised for resolving this difficulty should it occur. Actually it has never yet occurred in any practical application of linear programming, but artificial examples have been constructed to show that it can occur. Further discussion of degeneracy procedures is beyond the scope of this book (see Reference 8, Chapter 7).

Exercises

1. A school store sells slide rules at a profit of 50¢ each and sweatshirts at a profit of 40¢ each. It takes two minutes of a salesgirl's time and two minutes of a cashier's time to sell a slide rule. It requires three minutes of a salesgirl's time but only one minute of a cashier's time to sell a sweatshirt. The school store operates for a maximum of two hours during a school day, and during this time there is one cashier and there are two salesgirls available for handling sales of the above items. How many of each item should the store attempt to sell each day in order to earn a maximum profit?

(*Answer:* 30, 60; $39.)

2. A firm manufactures two types of machine screws (A and B) and sells them at a profit of 3¢ on type A and 4¢ on type B. Each type is processed on two machines, an automatic screw machine and a slotting machine. Type A screws require two minutes of processing time on the "automatic" and 5 minutes of processing time on the "slotter." Type B screws require three minutes on the "automatic" and two minutes on the "slotter." Each machine is available for not more than 60 hours during any working week. How many of each type of machine screw should the firm produce each week, in order to make a maximum profit?

(*Answer:* 327, 982; $49.09.)

3. A boat manufacturer builds both rowboats and canoes. Those built during the Spring and Summer go on sale in the Fall and Winter at a profit of \$20 per rowboat and \$18 per canoe. Those built during the Fall and Winter months, go on sale in the Spring and Summer at a profit of \$40 per rowboat and \$35 per canoe. Each rowboat requires 5 hours in the carpentry shops and 3 hours in the finishing shops. Each canoe requires 6 hours in the carpentry shops and 1 hour in the finishing shops. During each half year period, there are a maximum of 12,000 hours available in the carpentry shops and 15,000 hours available in the finishing shops. Enough materials are available to the manufacturer to build no more than 3000 rowboats and no more than 3000 canoes each year. How many of each kind of craft should he build during the Spring-Summer season, and how many during the Fall-Winter season, in order to realize a maximum profit?

(*Answer:* Rowboats: 600 during Spring-Summer
2400 during Fall-Winter
Canoes: 1500 during Spring-Summer
NONE during Fall-Winter
Maximum Profit: \$135,000.)

4. Taper pins are processed in a certain factory on two lathes, A and B, and also on a grinder G. The taper pins come in four sizes, identified as #1, #2, #6, and #8. The following table exhibits the processing time in hours per load, required on each machine, for each of the four types of pins. It also lists the net profit per load for each type and the maximum time available on each of the machines, weekly.

Pin Type	Lathe A	Lathe B	Grinder	Profit
#1	10	6	4.5	\$9
#2	5	6	18.0	7
#6	2	2	1.5	6
#8	1	2	6.0	4
Maximum hours available	50	36	81	

How many loads of each size taper pin should the factory produce in order to realize a maximum profit for the week's operation?

(*Answer:* Produce *only* #6 pins (18 loads) and NONE of the other sizes, for a maximum profit of \$108!)

NOTE. Problem 4 illustrates very nicely the fact that it is not always the best procedure to use the "most negative" last row entry to locate the variable to be introduced into the basis. Using this criterion it will be seen that three iterations are required to go from the initial B.F.S. to the optimal B.F.S. In the course of these iterations one of the slack variables, initially in the basis, will be seen to leave the basis only to return again and displace the variable which took its place. On the other hand, by choosing a suitable one of the other last row entries, the optimal

B.F.S. is reached in only one iteration! This interesting fact was discovered by one of my students.

The problems below require one or more artificial variables.

5. Mr. Phil N. Thrope wishes to spend up to \$120 for the purpose of buying Christmas toys for orphan children in a nearby home. He can buy dolls for \$3 each, skates at \$2 a pair, and toy rockets for \$5 each. He cannot buy more than 18 of the rockets and pairs of skates combined, but he must buy at least 20 of the pairs of skates and dolls combined. Determine (*a*) the *largest* number of toys, (*b*) the *smallest* number of toys, subject to the above restrictions.

(*Answer:* (*a*) 46 toys consisting of: 28 dolls
18 pairs of skates
0 rockets.
(*b*) 20 toys consisting of: 2 dolls
18 pairs of skates
0 rockets.)

6. Dr. Sy Koso Ma Teek, the eminent medical specialist, claims that he can cure colds with his special three-layer pills. These come in two sizes: the regular size, which contains 2 grains of aspirin, 5 grains of bicarbonate, and 1 grain of codeine; the king size, which contains 1 grain of aspirin, 8 grains of bicarbonate, and 6 grains of codiene. Now Dr. Sy Koso's researches have convinced him that it requires at *least* 12 grains of aspirin, 74 grains of bicarbonate, and 24 grains of codeine to effect the cure. Determine the least number of pills he should prescribe in order to meet these requirements.

(*Answer:* 10 pills consisting of 2 regular and 8 king size.)

7.* Daly Newsands, a student at the High School of Psyance, has two and a half idle hours on his hands, one day. The only things he can think of doing are against school regulations. Daly has a long dossier on file with the Dean, and he is under the constant surveillance of the discipline committee. There are various possible nefarious activities open to Daly, and he is particularly attracted by the following:

(*a*) Tacking up false information on the bulletin boards
(*b*) Making noise in the study hall
(*c*) Being in the lunchroom during the wrong period
(*d*) Walking on the left side of the corridor during change of subject periods

Daly is a realist, and he accepts stoically the fact that for every minute spent at activities (*a*), (*b*), (*c*), and (*d*), the discipline committee imposes a punishment of 6 days, 3 days, 4 days, and 1 day, respectively, in the detention hall. However, he is very hungry, and he cannot wait for his regular lunch period, so he decides to spend at least 15 illegal minutes in the lunchroom. He is a natural troublemaker and must spend at least 25 minutes misleading his fellow students by tampering with the bulletin

* Exercise 7 was constructed by several of my students.

boards. His program provides at most one hour which he can spend in the study hall and provides only four 5-minute changes of period. How shall Daly allocate his two and one-half idle hours among the various misdemeanors so as to suffer a MINIMUM amount of detention?

(*Answer:* 25, 60, 45, 20 (minutes) respectively, for a total of 530 days of detention.)

8. The Pension Board of a certain Board of Education invests the funds of its contributers in bonds, of which there are six forms of investment available. These are city bonds, labeled A_1 for preferred and A_2 for common; government bonds, labeled B_1 for preferred and B_2 for common; speculative bonds, labeled C_1 for preferred and C_2 for common. The yield from each type of investment is

Type:	A_1	A_2	B_1	B_2	C_1	C_2
Yield:	3%	$2\frac{1}{2}$%	$3\frac{1}{2}$%	4%	5%	$4\frac{1}{2}$%

The Board is restricted in its investment policy, by law, in the following way: (*a*) Investments in bonds of type A must not be less than 40% of total investment, and (*b*) Investments in bonds of types B and C shall not be more than 35% of total investment, *for either type.* Determine the investment program which will bring the maximum yield.

(*Answer:* 40% in type A_1 bonds, NONE in type A_2
25% in type B_2 bonds, NONE in type B_1
35 % in type C_1 bonds, NONE in type C_2
This brings the maximum yield of 3.95%.)

PART FOUR

Elementary Aspects of the Theory of Games

1. Preliminary Remarks

Among the more spectacular achievements of linear programming has been the light which it shed on a slightly older study, the Theory of Games of Strategy. This theory was first proposed in 1921 by the great French mathematician Emile Borel. It was successfully analyzed by John von Neumann, who proved its key result, the Minimax Theorem, in 1928. Together with Oskar Morganstern, von Neumann developed Game Theory as a method for analyzing competitive situations in economics, warfare, and other areas of conflicting interest. Their work was published in 1944, just about the time that linear programming came upon the scene. It was then recognized that problems in the Theory of Matrix Games could be formulated as special cases of linear programming. Thereupon,

Dantzig's Simplex Method became an important tool both in the practical and theoretical investigations of Game Theory. One of its triumphs was the elegant algebraic proof it provided for the Minimax Theorem.

In this part of our monograph we shall become acquainted with the elements of Game Theory. In Part Five we shall study how linear programming and the Simplex Method lead to an elegant solution of problems arising in Game Theory. Our discussion will be limited to the so-called "zero-sum, two-person games." These are also called *Matrix Games.*

2. 2 × 2 Matrix Games

Possibly the simplest and best known matrix game is matching pennies. In this game, two players A and B toss a penny independently of the other, and the outcomes are compared. If these outcomes "match," i.e., if they are both HEADS or both TAILS, then player A wins a penny from player B. If the outcomes do not match, i.e., if one player plays HEADS while the other plays TAILS, then B wins a penny from A. This game is neatly summarized in the following table:

		Player B	
		H	T
Player A	H	1	−1
	T	−1	1

The array of numerical entries $\begin{bmatrix} 1 & -1 \\ -1 & 1 \end{bmatrix}$ is called the *payoff matrix.* These entries represent the *amounts paid to player A*, for each of the four possible outcomes (H, H), (H, T), (T, H), (T, T). A negative entry therefore represents an actual payoff to B.

Now the device of tossing their respective coins independently of each other, at each "play" of the game, is merely a convenient way of insuring each player of two things,

(*a*) that he will not reveal to his opponent, in advance, the outcome of any particular play;

(*b*) that he will play HEADS or TAILS with *equal frequency*, "in the long run."

Item (*a*) is obviously important because either player could take immediate advantage of any such advance knowledge, by playing the side of his coin which would produce a win for him. Some form of "randomization" is, therefore, imperative to prevent either player from knowing in advance what the next outcome will be. On the other hand, the importance of item (*b*) is not immediately obvious, because playing HEADS or TAILS with equal frequency is only one of many possible "strategies" which either player could adopt. Each player could vary his strategy by varying the relative frequency with which he plays HEADS or TAILS. For example, *A* might play HEADS two-thirds of the time and TAILS only one-third of the time, while *B* might play HEADS one-fourth of the time, with TAILS therefore coming up three-fourths of the time. This could all be accomplished in a completely random manner by providing each player with a card on which is mounted a pointer which he can spin over a suitably calibrated dial.

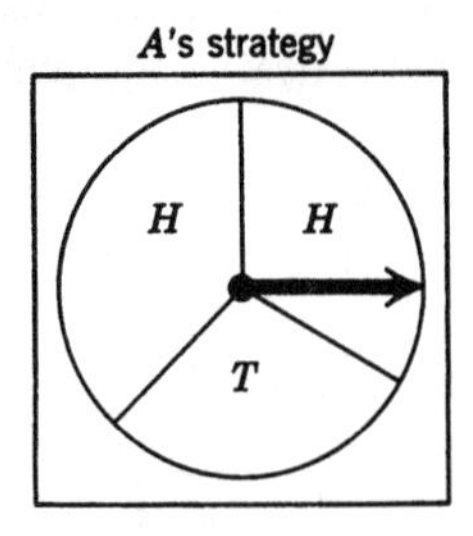

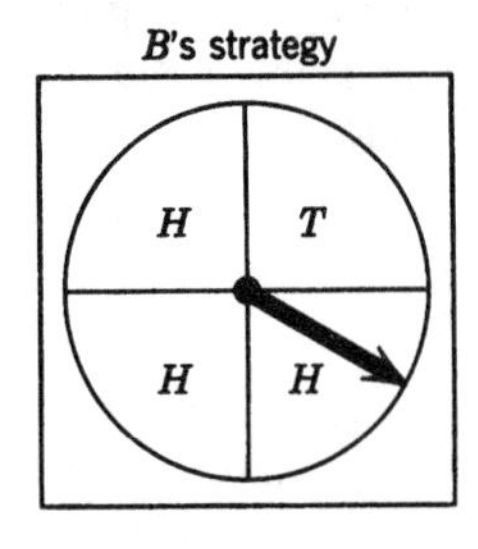

Each of the *ordered pairs* of frequencies (probabilities), such as $(\frac{1}{2}, \frac{1}{2})$, $(\frac{2}{3}, \frac{1}{3})$, $(\frac{1}{4}, \frac{3}{4})$, $(1, 0)$, etc. is termed a *strategy* in Game Theory. Clearly, in this context any ordered pair of real numbers (x, y), where $x \geq 0$, $y \geq 0$ and $x + y = 1$, can represent a possible strategy for consideration.

It is desirable to distinguish among the various possible versions of a matching pennies game, which can result from changes of strategy by either or both players. We can do this by inserting the particular strategy used by each player beside the payoff matrix. For example we can represent the ordinary version of matching pennies, where both players employ the equal frequency strategy $(\frac{1}{2}, \frac{1}{2})$, as follows:

Player B

		H	T
		$\frac{1}{2}$	$\frac{1}{2}$
Player A: H	$\frac{1}{2}$	1	−1
T	$\frac{1}{2}$	−1	1

(Ordinary version of matching pennies with both players using the "equal frequency" strategy)

For the modified version of the game, where A uses the strategy $(\frac{2}{3}, \frac{1}{3})$ while B uses the strategy $(\frac{1}{4}, \frac{3}{4})$, we have the following diagram:

Player B

		H	T
		$\frac{1}{4}$	$\frac{3}{4}$
Player A: H	$\frac{2}{3}$	1	−1
T	$\frac{1}{3}$	−1	1

(Modified version of matching pennies, with each player using a different strategy, neither using "equal frequency")

Let us now compare these two versions of the game from the point of view of each of the players. If the ordinary version of the game is repeated a great many times, then on half of the occasions that A plays HEADS, he can expect to win a penny while on half of these occasions he can expect to lose a penny. Similarly, on half of the occasions that A plays TAILS, he can expect to win while on half of these occasions he should expect to lose. This is, of course, due to the fact that B is *independently* playing HEADS or TAILS with equal frequency. By so doing, B is making certain that A's (long run) *expectation* is zero, regardless of how frequently A plays HEADS or TAILS. Thus on repeated plays of the ordinary matching pennies game A's *average winnings per play* may be computed as follows:

(1) A plays HEADS one-half of the time; hence, because of B's equal frequency strategy, A wins (on these occasions) an average of

$$\tfrac{1}{2}[\tfrac{1}{2}(1) + \tfrac{1}{2}(-1)] = \tfrac{1}{2}(0) = 0 \quad \textit{per play}.$$

(2) A plays TAILS one-half of the time; hence, because of B's equal frequency strategy, A wins (on these occasions) an average of

$$\tfrac{1}{2}[\tfrac{1}{2}(-1) + \tfrac{1}{2}(1)] = \tfrac{1}{2}(0) = 0 \quad \textit{per play}.$$

(3) Adding the amounts calculated in (1) and (2), we see that A's average winnings per play = 0.

Clearly, if A used any other strategy (x, y) instead of $(\tfrac{1}{2}, \tfrac{1}{2})$, in this game, then his average winnings per play would still be 0, because

$$x[\tfrac{1}{2}(1) + \tfrac{1}{2}(-1)] + y[\tfrac{1}{2}(-1) + \tfrac{1}{2}(1)] = x[0] + y[0] = 0.$$

We repeat: this is due to the equal frequency strategy $(\tfrac{1}{2}, \tfrac{1}{2})$ which B is using.

By symmetry, it is equally clear that so long as A uses the strategy $(\tfrac{1}{2}, \tfrac{1}{2})$, he makes certain that B's (long run) expectation is zero, regardless of B's strategy. Suppose, however that A adopts a strategy (x, y) other than $(\tfrac{1}{2}, \tfrac{1}{2})$; let us say $x = \tfrac{2}{3}$, $y = \tfrac{1}{3}$. Then by using an "opposite" strategy, let us say $(\tfrac{1}{4}, \tfrac{3}{4})$, player B can inflict a long run loss on A. In fact this particular pair of strategies, namely $(\tfrac{2}{3}, \tfrac{1}{3})$ for A and $(\tfrac{1}{4}, \tfrac{3}{4})$ for B, yields the modified version of the game which we diagrammed above. We can compute A's average winnings per play, for this version of the game as follows:

(1) A plays HEADS two-thirds of the time; on one-fourth of these occasions B plays HEADS while on three-fourths of these occasions B plays TAILS; hence A's average winnings for these occasions are

$$(\tfrac{2}{3})[(\tfrac{1}{4})(1) + (\tfrac{3}{4})(-1)] = (\tfrac{2}{3})[-\tfrac{1}{2}] = -\tfrac{1}{3} \quad \textit{per play}.$$

(2) A plays TAILS one-third of the time; on one-fourth of these occasions B plays HEADS while on three-fourths of these occasions B plays TAILS; hence A's average winnings for these occasions are

$$(\tfrac{1}{3})[(\tfrac{1}{4})(-1) + (\tfrac{3}{4})(1)] = (\tfrac{1}{3})[\tfrac{1}{2}] = \tfrac{1}{6} \quad \textit{per play}.$$

(3) Adding these two amounts, we find that A's average winnings per play are $(-\tfrac{1}{3}) + (\tfrac{1}{6}) = -\tfrac{1}{6}$ *per play*.

This result means that in the modified game A actually *loses* to B an average of $\tfrac{1}{6}$¢ per play (in the long run). It is, of course, obvious that B *wins* on the average $\tfrac{1}{6}$¢ per play, but this fact can also be verified independently by analyzing the game from B's point of view. We shall make this direct analysis this one time:

(1) B plays HEADS $\frac{1}{4}$ of the time. On $\frac{2}{3}$ of these occasions A plays HEADS while on $\frac{1}{3}$ of these occasions A plays TAILS; hence B's average *losses* for these occasions amount to

$$(\tfrac{1}{4})[(\tfrac{2}{3})(1) + (\tfrac{1}{3})(-1)] = (\tfrac{1}{4})[\tfrac{1}{3}] = \tfrac{1}{12} \quad \textit{per play.}$$

(2) B plays TAILS $\frac{3}{4}$ of the time. On $\frac{2}{3}$ of these occasions A plays HEADS while on $\frac{1}{3}$ of these occasions A plays TAILS; hence B's average *losses* for these occasions amount to

$$(\tfrac{3}{4})[(\tfrac{2}{3})(-1) + (\tfrac{1}{3})(1)] = (\tfrac{3}{4})[-\tfrac{1}{3}] = -\tfrac{1}{4} \quad \textit{per play.}$$

(3) Adding the two amounts calculated in (1) and (2), we see that B's average *losses* per play amount to $\frac{1}{12} - \frac{1}{4} = -\frac{1}{6}$.

A *loss* (to B) of $-\frac{1}{6}$ is of course a *gain* (to B) of $\frac{1}{6}$. This result emphasizes the point that all payoffs calculated from the payoff matrix represent amounts paid to player A. When these amounts are negative, they represent actual gains for player B.

It is now fairly clear that any version of the matching pennies game, in which one player decides to adopt a strategy other than $(\frac{1}{2}, \frac{1}{2})$, will turn out to be unfavorable to that player (in the long run), assuming that his opponent is clever enough to recognize what is happening and to take advantage of it. It is customary in Game Theory to suppose that the two players are equally clever. One can then prove on this basis that $(\frac{1}{2}, \frac{1}{2})$ is indeed the "optimal" strategy for both players. In fact suppose that A adopts the strategy $(x, 1 - x)$ while B adopts the strategy $(y, 1 - y)$:

			Player B	
			H	T
			y	$1 - y$
Player A	H	x	1	-1
	T	$1 - x$	-1	1

In repeated plays of this version of the game, A can expect an average gain z, per play, given by

$$z = x[y(1) + (1 - y)(-1)] + (1 - x)[y(-1) + (1 - y)(1)].$$

This readily reduces to

$$z = 4xy - 2x - 2y + 1,$$

or equivalently,

$$z = (2x - 1)(2y - 1) = 4(x - \tfrac{1}{2})(y - \tfrac{1}{2}).$$

B's expected gain per play is, of course, $-z$. A desires to *maximize* z, while B desires to maximize $-z$, i.e., to *minimize* z. Now if either $x = \frac{1}{2}$ or $y = \frac{1}{2}$, then $z = 0$, indicating neither gain nor loss for either player, provided at least one of them chooses the strategy $(\frac{1}{2}, \frac{1}{2})$. On the other hand, if say $x > \frac{1}{2}$, then the factor $(x - \frac{1}{2})$ is *positive*, so by choosing $y < \frac{1}{2}$ player B can make the second factor $(y - \frac{1}{2})$ *negative*, thereby making the value of z *negative*. In fact, by choosing the strategy $(0, 1)$, player B can make A's gain amount to $-(2x - 1)$. This is clearly B's best strategy whenever A chooses a strategy $(x, 1 - x)$ with $x > \frac{1}{2}$. Similarly, if A chooses a strategy with $x < \frac{1}{2}$, then B's best strategy is $(1, 0)$.

The particular strategies $(1, 0)$ and $(0, 1)$ are also called *pure strategies.* A player using a pure strategy will play exclusively HEADS or exclusively TAILS. All other strategies, such as $(\frac{1}{2}, \frac{1}{2})$, $(\frac{3}{4}, \frac{1}{4})$, etc. are called *mixed strategies.*

Returning to the analysis above, it is obvious that unless player A is unusually stupid, he will not adhere very long to any strategy that will enable player B to force upon him a long run loss. The only strategy that prevents this is the mixed strategy $(\frac{1}{2}, \frac{1}{2})$. By using it, A insures that his long run gain z, cannot be brought below 0. On the other hand this strategy makes $z = 0$. This is therefore the *maximum* long run gain which A can hope to achieve against a clever opponent! In this sense, $(\frac{1}{2}, \frac{1}{2})$ is A's *optimal strategy.* Similarly, player B minimizes z by adopting the optimal mixed strategy $(\frac{1}{2}, \frac{1}{2})$, i.e., this strategy is optimal for B as well as for A. B's minimum long run loss is also 0. This state of affairs is described by saying that 0 is the *value* of the game and a game whose value is 0 is called *fair.*

It is very illuminating to draw a (three dimensional) graph of the equality

$$z = 4xy - 2x - 2y + 1.$$

This graph is shown in Figure 1. It is known technically as a *hyperbolic paraboloid*; more commonly as a *SADDLE SURFACE.* The *saddle point* O' of this surface is located in the XY-plane, and its coordinates are $(\frac{1}{2}, \frac{1}{2}, 0)$. The curve $O'A$ is a parabola curving downward *below* the XY-plane and having its *highest* point at O'. The curve $O'B$ is a parabola curving upward *above* the XY-plane and having its *lowest* point at $0'$. The value of z at point O' is zero and because this value is simultaneously a maximum along curve

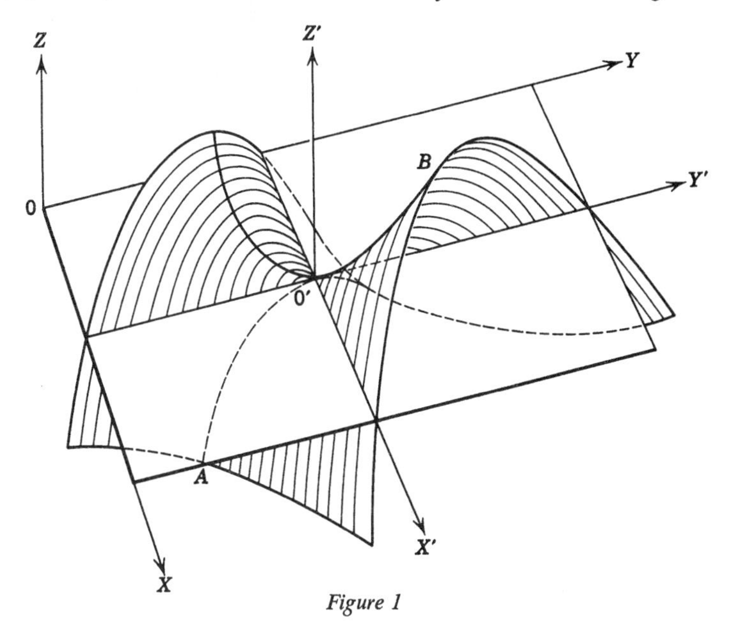

Figure 1

$O'A$ and a minimum along curve $O'B$, it is called a *minimax* value of z. We shall see that this sort of phenomenon is characteristic of all matrix games, not merely the simple matching pennies game which we have thus far been discussing. However, we shall continue a bit further with this simple 2 × 2 matrix game.

Let us examine the effect on our matching pennies game of altering the payoff matrix. Consider for example the following payoff matrix:

		Player *B*	
		H	*T*
Player *A*	*H*	1	−2
	T	−3	4

In this new game player A still wins when the coins match, and player B wins when the coins do not match. However the payoffs are no longer the same amounts in all cases. This time A stands to win either 1¢ or 4¢, and he stands to lose either 3¢ or 2¢. Superficially, it appears that A's gains should offset A's losses in a long run sequence of plays, so it appears to be a "fair" game. Let us examine whether this is really so.

If the game is played in the usual way, i.e., by tossing the coins before each play, then both A and B will be using the same mixed strategy, namely, $(\frac{1}{2}, \frac{1}{2})$. A's expectation (long run gain per play) is readily calculated:

$$z = \tfrac{1}{2}[\tfrac{1}{2}(1) + \tfrac{1}{2}(-2)] + \tfrac{1}{2}[\tfrac{1}{2}(-3) + \tfrac{1}{2}(4)] = -\tfrac{1}{4} + \tfrac{1}{4} = 0.$$

B's expected long run *loss* per play must therefore also be 0. We verify this directly as follows:

$$z = \tfrac{1}{2}[\tfrac{1}{2}(1) + \tfrac{1}{2}(-3)] + \tfrac{1}{2}[\tfrac{1}{2}(-2) + \tfrac{1}{2}(4)] = -\tfrac{1}{2} + \tfrac{1}{2} = 0.$$

All this seems fair enough! But remember that we are supposing that A is very clever (and so is B). A will ask himself whether $(\frac{1}{2}, \frac{1}{2})$ is really his optimal strategy. After all, if B were to continue to play HEADS or TAILS with equal frequency, then A could switch to the pure strategy $(0, 1)$ and his expected gain per play would become:

$$z = 0[\tfrac{1}{2}(1) + \tfrac{1}{2}(-2)] + 1[\tfrac{1}{2}(-3) + \tfrac{1}{2}(4)] = 0 + \tfrac{1}{2} = \tfrac{1}{2},$$

i.e., player A could now expect to win ½¢ per play. But B is no fool either, and he would not permit this to go on. He could, for example switch to the "opposite" pure strategy $(1, 0)$ and his expected *loss* (A's gain) per play would become:

$$z = 1[0(1) + 1(-3)] + 0[0(-2) + 1(4)] = -3 + 0 = -3,$$

which means that B would win 3¢ per play from A, who would now of course, quickly seek a new change of strategy!

This "hit and miss" analysis is readily replaced by a more systematic mathematical approach. As in the previous game, let us suppose that A chooses the strategy (x, y), and B chooses the strategy (u, v). A's expected gain per play becomes

$$\begin{aligned} z &= x[u(1) + v(-2)] + y[u(-3) + v(4)] \\ &= x[u - 2(1 - u)] + (1 - x)[-3u + 4(1 - u)] \\ &= x(3u - 2) + (1 - x)(4 - 7u) \\ &= 3xu - 2x + 4 - 7u - 4x + 7xu \\ &= 10xu - 6x - 7u + 4 \\ &= 10(xu - \tfrac{6}{10}x - \tfrac{7}{10}u) + 4 \\ &= 10(x - \tfrac{7}{10})(u - \tfrac{6}{10}) - \tfrac{42}{100} + 4 \\ &= 10(x - \tfrac{7}{10})(u - \tfrac{6}{10}) - 4.2 + 4 \\ &= 10(x - \tfrac{7}{10})(u - \tfrac{6}{10}) - .2. \end{aligned}$$

The true nature of this game is now apparent! For, if B chooses $u = \frac{6}{10}$, then regardless of the value of x, the first term on the right vanishes and z attains the fixed value $-.2$. In other words, the strategy $(\frac{6}{10}, \frac{4}{10})$, if adopted by B, forces A to lose to B an average of $\frac{1}{5}$¢ per play, regardless of what strategy A may adopt. Furthermore A had better choose the strategy $(\frac{7}{10}, \frac{3}{10})$, i.e., he should choose $x = \frac{7}{10}$. If he does not do this, say if $x < \frac{7}{10}$, then B can choose $u > \frac{6}{10}$. This will make the first term negative, thereby augmenting A's loss. Similarly, if A chose $x > \frac{7}{10}$, then B could choose $u < \frac{6}{10}$ and once again win from A even more than $\frac{1}{5}$¢ per play. We leave it to the reader to verify that the strategy $(\frac{6}{10}, \frac{4}{10})$ IS LIKEWISE FORCED UPON B. These two particular mixed strategies $(\frac{7}{10}, \frac{3}{10})$ for A, and $(\frac{6}{10}, \frac{4}{10})$ for B, are the *optimal* strategies for this game. The value of this game is $-\frac{1}{5}$¢. The value represents A's maximum gain and B's minimum loss. Clearly *the game is not fair to A*! However it can be made fair by arranging that B shall pay A one cent in advance, for every five plays of the game.

As another illustration of our method for analyzing a 2 × 2 matrix game, consider the following payoff matrix:

Player B

Player A		H	T
	H	-2	-4
	T	3	5

If A chooses a strategy (x, y) and B chooses a strategy (u, v) then A's expected gain per play becomes

$$\begin{aligned} z &= x(-2u - 4v) + y(3u + 5v) \\ &= x[-2u - 4(1 - u)] + (1 - x)[3u + 5(1 - u)] \\ &= x(2u - 4) + (1 - x)(-2u + 5) \\ &= 4xu - 9x - 2u + 5 \\ &= 4(xu - \tfrac{9}{4}x - \tfrac{1}{2}u) + 5 \\ &= 4[(x - \tfrac{1}{2})(u - \tfrac{9}{4}) - \tfrac{9}{8}] + 5 \\ &= 4(x - \tfrac{1}{2})(u - \tfrac{9}{4}) + \tfrac{1}{2}. \end{aligned}$$

Player A controls the value of the first factor $(x - \frac{1}{2})$, while B controls the second factor $(u - \frac{9}{4})$. But, as $0 \leq u \leq 1$, the second factor $(u - \frac{9}{4})$ will always be negative, regardless of which strategy is chosen by B. Therefore, A will certainly choose $x < \frac{1}{2}$ so as to make the first factor negative also. In fact A will try to make the product of the two factors as large as possible, while B will try to make this product as small as possible. It is evident that A should choose $x = 0$, while B should choose $u = 1$. The optimal strategies are therefore *pure strategies*:

$$A\colon\ (0, 1) \quad \text{and} \quad B\colon\ (1, 0).$$

The value of the game is

$$v = 4(0 - \tfrac{1}{2})(1 - \tfrac{9}{4}) + \tfrac{1}{2} = 4(-\tfrac{1}{2})(-\tfrac{5}{4}) + \tfrac{1}{2},$$

i.e., $$v = 3.$$

A game with optimal pure strategies is also called *strictly determined.*

The existence of optimal pure strategies for both A and B could have been predicted in advance by the following "minimax" argument: The *minimum* payoff in row 1 is -4, while the *minimum* payoff in row 2 is 3. These are the "worst" A can expect from each of his two available pure strategies. Naturally, he prefers the *maximum* of these two minimum payoffs, and he realizes that adoption of the pure strategy (0, 1) will assure him *at least* this amount. Player B, on the other hand, observes that the *maximum* payoff (to A) in column 1 is 3, while the *maximum* payoff to A in column 2 is 5. These are the "worst" B can expect from each of his own pure strategies. Naturally, he prefers the *minimum* of these two maximum payoffs to A, and he realizes that adoption of the pure strategy (1, 0) by him, will prevent A from getting any more than this amount.

The presence in the payoff matrix of an entry such as 3, which is simultaneously a maximum of row minimums as well as a minimum of column maximums, is characteristic of strictly determined games. Such games are said to have a *saddle point*. In the present illustration this *minimax* entry appears at the intersection of row 2 and column 1 of the payoff matrix. The ordered pair of indices (2, 1) is called a *saddle point*, of this matrix. In general, if a payoff matrix possesses a minimax entry in row i and column j, then the ordered pair (i, j) is called a *saddle point* of the matrix. The game is then strictly determined, i.e., it has optimal pure strategies for both players, and the value of the game is the same as the value at the saddle point.

While strictly determined games are especially easy to solve in the manner just described, observe that they can also be handled by the same algebraic analysis which we used on nonstrictly determined games. This analysis works very well for 2×2 matrix games, but for more elaborate games it becomes too unwieldy. In Section 4, we shall see that linear programming with its ingenious Simplex Method provides a technique powerful enough to cope with more general rectangular games. Meanwhile, we shall study in the next section some interesting graphical techniques capable of handling games with a $2 \times n$ or $n \times 2$ payoff matrix.

The following exercises yield further insight into 2×2 matrix games.

Exercises

1. In the following 2×2 matrix games, player A chooses rows and player B chooses columns. Determine optimal strategies for both players and the value of each game.

(*a*) $\begin{bmatrix} 0 & 3 \\ 2 & 1 \end{bmatrix}$ (*b*) $\begin{bmatrix} -1 & 2 \\ 1 & 0 \end{bmatrix}$ (*c*) $\begin{bmatrix} 4 & -3 \\ 0 & 2 \end{bmatrix}$

(*d*) $\begin{bmatrix} -2 & 3 \\ 3 & -4 \end{bmatrix}$ (*e*) $\begin{bmatrix} -2 & -4 \\ 3 & 5 \end{bmatrix}$ (*f*) $\begin{bmatrix} 0 & -3 \\ 3 & 0 \end{bmatrix}$

(*g*) $\begin{bmatrix} -4 & 3 \\ 6 & 0 \end{bmatrix}$

2. Given the 2×2 payoff matrix $\begin{bmatrix} a & b \\ c & d \end{bmatrix}$ let player A adopt the strategy (x, y), while player B adopts the strategy (u, v), where x, y, u, v, are all ≥ 0 and $x + y = u + v = 1$.

a. Express A's expected gain z, in terms of x, y, u, v, a, b, c, and d.

b. What is the effect on z of adding the same constant k to each number in the payoff matrix?

c. What is the effect on z of multiplying each payoff entry by the same constant k?

d. How are the optimal strategies affected by these operations on the payoff matrix?

3. Show that the 2×2 matrix game $\begin{bmatrix} a & a \\ c & d \end{bmatrix}$ is strictly determined, i.e., show that if two entries in a row are equal, then the matrix has a saddle point. Similarly, for columns.

4. Prove that the 2×2 matrix game $\begin{bmatrix} a & b \\ c & d \end{bmatrix}$ is nonstrictly determined, if and only if

either: $\quad a < b, \quad a < c, \quad d < b, \quad \text{and} \quad d < c,$

or: $\quad a > b, \quad a > c, \quad d > b, \quad \text{and} \quad d > c,$

5. Determine optimal strategies and the value for the game defined by the payoff matrix $\begin{bmatrix} a & 0 \\ 0 & b \end{bmatrix}$.

6. Given the 2×2 matrix game defined by $\begin{bmatrix} a & -b \\ -c & d \end{bmatrix}$ where a, b, c, and d, are all ≥ 0. Prove that the optimal strategies are:

$$A: \left(\frac{c + d}{a + b + c + d}, \frac{a + b}{a + b + c + d}\right)$$

$$B: \left(\frac{b + d}{a + b + c + d}, \frac{a + c}{a + b + c + d}\right),$$

and the value of the game is

$$v = \frac{ad - bc}{a + b + c + d}.$$

(*Note:* This shows that a 2×2 matrix game having the prescribed form is fair, if and only if $ad - bc = 0$. This leads to the somewhat surprising conclusion that a game such as $\begin{bmatrix} 2 & -3 \\ -3 & 4 \end{bmatrix}$ is *not fair* despite the fact A's payoffs add up to the same total as B's payoffs.)

(*Answers to Exercises 1 and 5;*)

1. (*a*) A: $(\frac{1}{4}, \frac{3}{4})$, B: $(\frac{1}{2}, \frac{1}{2})$, $v = \frac{3}{2}$.
   (*b*) A: $(\frac{1}{4}, \frac{3}{4})$, B: $(\frac{1}{2}, \frac{1}{2})$, $v = \frac{1}{2}$.
   (*c*) A: $(\frac{2}{9}, \frac{7}{9})$, B: $(\frac{5}{9}, \frac{4}{9})$, $v = \frac{8}{9}$.
   (*d*) A: $(\frac{7}{12}, \frac{5}{12})$ B: $(\frac{7}{12}, \frac{5}{12})$, $v = \frac{1}{12}$.
   (*e*) A: $(0, 1)$, B: $(1, 0)$, $v = 3$.
   (*f*) A: $(0, 1)$, B: $(0, 1)$, $v = 0$.
   (*g*) A: $(\frac{6}{13}, \frac{7}{13})$, B: $(\frac{3}{13}, \frac{10}{13})$, $v = \frac{18}{13}$

5. $\left(\frac{b}{a+b}, \frac{a}{a+b}\right)$ for both players $v = \frac{ab}{a+b}$.

3. Graphical Analysis of m x 2 and 2 x n Matrix Games

Consider the matrix game

$$\text{Player } A \quad \overset{\text{Player } B}{\begin{bmatrix} 4 & -3 \\ 0 & 2 \end{bmatrix}}$$

The rows of this matrix are ordered pairs of real numbers

$$P = (4, -3),$$
$$Q = (0, \quad 2).$$

Let us plot these as points on a graph. (See Figure 2. The other points and lines of this figure will be explained as we proceed.)

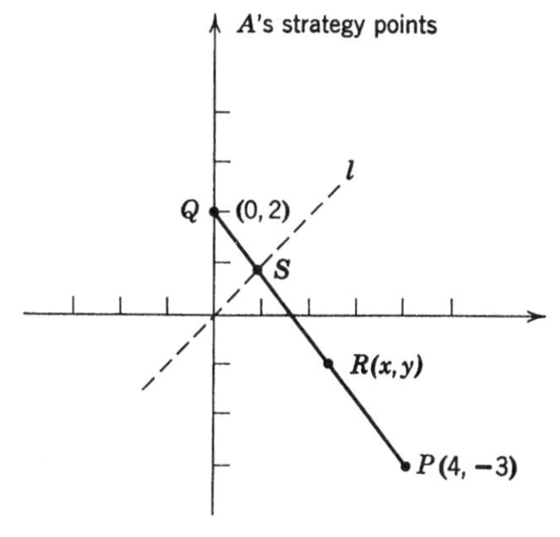

Figure 2

Suppose now, that player A adopts a strategy $(t, 1 - t)$, where $0 \leq t \leq 1$. Then, against B's first pure strategy $(1, 0)$, A's expected payoff (we shall call it x) is given by

$$x = t(4) + (1 - t)(0), \quad \text{i.e.,} \quad x = 0 + 4t.$$

Against B's second pure strategy $(0, 1)$, A's expected payoff will be

$$y = t(-3) + (1 - t)(2), \quad \text{i.e.,} \quad y = 2 - 5t.$$

As t varies from 0 to 1, we obtain all the points (x, y) of segment $\overset{\bullet\!-\!\bullet}{QP}$ (see discussion of parametric equations in Chapter 2). Consequently, there is a one-to-one correspondence between the points of this segment and the set of all possible (mixed) strategies available to A. Observe that this segment is a convex set. It is in fact the "smallest" convex set which contains both points P and Q. It is called the *convex hull* of P and Q.

Now, if A has selected a particular strategy, say one which corresponds to the point R on segment $\overset{\bullet\!-\!\bullet}{QP}$, then B will naturally prefer to select from among all his own available strategies, pure or mixed, one which minimizes the payoff to A. *This minimum payoff will always be the smaller of the two coordinates of R, x or y.* To prove this, suppose that B were to adopt the strategy (p, q), where $p + q = 1$. The resulting expected payoff to A will then be $px + qy$. Now if $x \leq y$, then

$$x = (p + q)x = px + qx \leq px + qy \leq py + qy \leq (p + q)y = y.$$

So *if* $x \leq y$, then $x \leq px + qy \leq y$.

Similarly, *if* $y \leq x$, then $y \leq px + qy \leq x$.

Either way, the minimum payoff, for all possible strategies (p, q) which B may adopt, is the *smaller* of the two coordinates of A's strategy point. This means that B can achieve maximum protection against each specified strategy of A pure or mixed, by using one or the other of B's *pure* strategies, namely the one corresponding to the smaller coordinate of A's stratehy point.

Suppose, therefore, that the line l, defined by $y = x$ be drawn on the diagram. If A's strategy point R, lies above this line l, then B should use his own *first* pure strategy, i.e. $(1, 0)$, because all points above line l have a smaller first coordinate. If A's strategy point R

lies below the line l, then B should adopt his own *second* pure strategy, i.e., (0, 1), because all points below line l have a smaller second coordinate. If point R happens to lie on line l, then it is immaterial which pure strategy B adopts, because the value of both coordinates is the same in such a case.

Knowing that B will seek to minimize the payoff, *A should, of course, select that strategy point of segment* $\overline{PQ}$, *whose smaller coordinate is as large as possible*! In the present example, this turns out to be the point S, where line l intersects segment $\overline{PQ}$, because all points of segment $\overline{QP}$ which are below l, have an ordinate smaller than that of S, while all points of the segment which are above l, have an abscissa smaller than the abscissa of S. The point S is readily found by equating x and y.

$$0 + 4t = 2 - 5t; \text{ therefore } 9t = 2, \text{ i.e., } t = \tfrac{2}{9}.$$

This yields $x = y = \frac{8}{9}$, which is the value of the game to A. His optimal strategy thus appears to be $(\frac{2}{9}, \frac{7}{9})$.

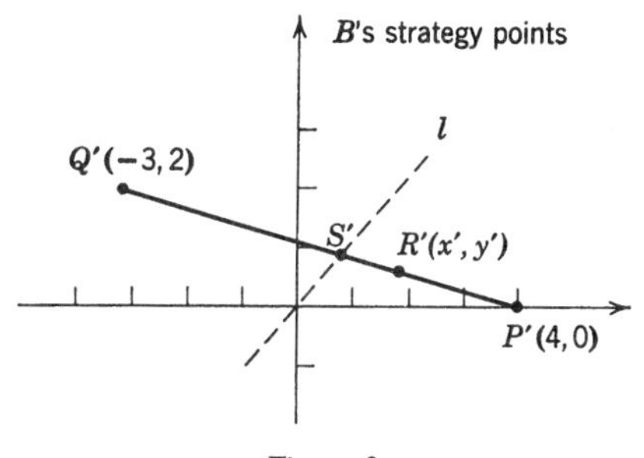

Figure 3

Let us next seek an optimal strategy for B. To do this we plot the *columns* of the matrix as ordered pairs of numbers. These are indicated in Figure 3 as

$$P' = (4, 0), \qquad Q' = (-3, 2).$$

Suppose player B adopts a strategy $(s, 1 - s)$. Then against A's first pure strategy, B must expect to pay off the amount

$$x' = s(4) + (1 - s)(-3), \qquad \text{i.e.,} \quad x' = -3 + 7s.$$

Against A's second pure strategy, he should expect to pay

$$y' = s(0) + (1 - s)(2), \qquad \text{i.e.,} \quad y' = 2 - 2s.$$

As s varies from 0 to 1, these parametric equations yield all the points of segment $\overline{Q'P'}$. The points of this segment are in one-to-one correspondence with the mixed strategies available to B. If B selects the strategy point $R' = (x', y')$, A will naturally prefer to select from among all his possible strategies, one which maximizes the payoff. This maximum payoff will clearly be the larger of the two coordinates x' and y'. If the point R' lies above the line l (defined by $y = x$), then A should clearly select his own second pure strategy $(0, 1)$, because the second coordinate of R' is larger than the first coordinate, for all points above l. If R' lies below l, then A should select his own first pure strategy $(1, 0)$. Should R' happen to fall upon l, then A's choice is immaterial.

Knowing that A will seek to maximize the payoff, *B should of course, select that point of segment* $\overline{Q'P'}$ *whose larger coordinate is as small as possible*! In the present example, this is the point S' where segment $\overline{Q'P'}$ intersects line l. This is true because all points of the segment which are above l, have an ordinate larger than S', while all points below l, have an abscissa larger than that of S'. We find S' by equating x' and y'.

$$-3 + 7s = 2 - 2s; \quad \text{therefore} \quad 9s = 5, \quad \text{i.e.,} \quad s = \tfrac{5}{9}.$$

This value of s yields $x' = y' = \frac{8}{9}$. This is the *most* that B need expect to pay off to A, provided B adopts the (optimal) strategy $(\frac{5}{9}, \frac{4}{9})$.

That A's minimum expected gain is the same as B's maximum expected loss is, of course, not accidental. This simple example is a particular illustration of the very fundamental Minimax Theorem, first proved by John von Neumann in 1928. However, before considering the general problem it will be illuminating to examine several further special cases.

Consider the matrix game:

$$\begin{array}{cc} & \textit{Player B} \\ \textit{Player A} & \begin{bmatrix} -2 & -4 \\ 3 & 5 \end{bmatrix}. \end{array}$$

Using the rows we plot the strategy set for A and using the columns we plot the strategy set for B. As in the previous example, these strategy sets are segments, each segment being the smallest convex set containing the two *pure* strategy points for each player. The strategy sets are depicted in Figures 4 and 5. In this game, A's optimum strategy point (i.e., the point whose smaller coordinate is as large as possible) does not lie at the intersection of segment $\overline{QP}$

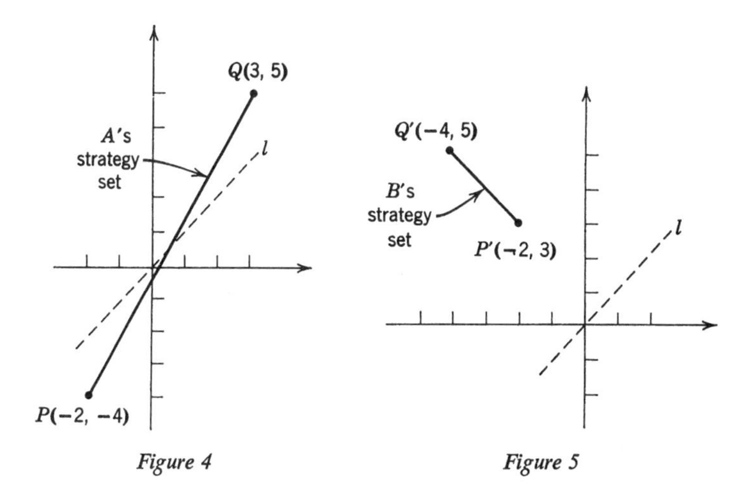

Figure 4

Figure 5

with the line l. It is, instead, the endpoint Q of the segment. Similarly, B's optimum strategy point (i.e., the one whose larger coordinate is as small as possible) is the point P' at the end of segment $\overline{Q'P'}$. This segment does not even intersect line l. In this game both A and B have optimal *pure* strategies (0, 1) and (1, 0), respectively. The expected payoff to player A is 3 in both cases, but this fact is less surprising than it was in the preceding example, because we have already pointed out in the previous section that this particular game is strictly determined. (We saw that a saddle point occurred at row 2, column 1, where the minimax value 3 appears in the matrix.) The really surprising development is that matrix games without a saddle point also have optimal strategies for both players and that these optimal strategies lead to exactly the same long run payoff which constitutes both a maximum for player A, as well as a minimum for player B.

The graphical procedure described here can be applied to $m \times 2$ or $2 \times n$ matrices even when $m \geq 3$ or $n \geq 3$. For example, consider the 3×2 matrix game.

$$\text{Player } A \quad \overset{\text{Player } B}{\begin{bmatrix} 2 & 5 \\ 3 & 1 \\ 0 & 3 \end{bmatrix}}$$

The rows of this matrix are plotted as points P, Q, and R (see Figure 6). A mixed strategy for A now consists of an ordered *triple* (p, q, r),

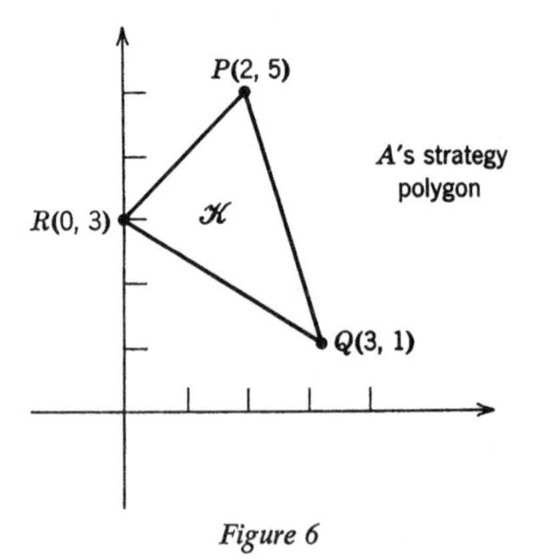

Figure 6

whose entries are non-negative numbers adding up to 1, i.e.,

$$p \geq 0, q \geq 0, r \geq 0 \quad \text{and} \quad p + q + r = 1.$$

Against B's first pure strategy (1, 0), A's expected payoff is

$$x = p(2) + q(3) + r(0).$$

Against B's second pure strategy (0, 1), A's expected payoff is

$$y = p(5) + q(1) + r(3).$$

The point with coordinates (x, y), is readily shown to lie within or on the boundary of a convex polygon, $\mathscr{K}$. This polygon (a triangle, in the present case) is formed by joining each pair of strategy points by a segment, then joining every pair of points in this strategy set by a segment once again, and so forth. This produces the smallest convex set (in the plane) which contains the original pure startegy points of player A. This strategy set is called the *convex hull* of the original points (P, Q, and R, in the present example).

To prove that the point (x, y) lies within or on the boundary of this convex polygon, we proceed as follows: let $q + r = s$; then $s \geq 0$ and $p + s = 1$. If $s = 0$, then $p = 1$ and (x, y) coincides with point P. Furthermore, if $s > 0$, then we write

$$x = p(2) + s\left[\frac{q}{s}(3) + \frac{r}{s}(0)\right] = p(2) + s(x_1),$$

$$y = p(5) + s\left[\frac{q}{s}(1) + \frac{r}{s}(3)\right] = p(5) + s(y_1),$$

where $\quad x_1 = \frac{q}{s}(3) + \frac{r}{s}(0) \quad$ and $\quad y_1 = \frac{q}{s}(1) + \frac{r}{s}(3).$

Now if we rewrite $\quad \frac{q}{s} = t, \quad$ then $\quad \frac{r}{s} = 1 - t.$ Hence,

$$x_1 = t(3) + (1 - t)(0),$$
$$y_1 = t(1) + (1 - t)(3),$$

where

$$0 \leq t \leq 1.$$

Therefore (x_1, y_1) is a point on segment $\overset{\bullet\!-\!\bullet}{QR}$. Call this point S. Observe next that because $s = 1 - p$, we can write

$$\begin{aligned} x &= p(2) + (1 - p)x_1 \\ y &= p(5) + (1 - p)y_1 \end{aligned} \qquad \text{(where } 0 \leq p \leq 1\text{)}.$$

This shows that (x, y) lies on segment $\overset{\bullet\!-\!\bullet}{PS}$ and therefore belongs either in the interior or on the boundary of triangle PQR (see Figure 7*a*).

An obvious reversal of this argument shows that, conversely, any point of this convex set $\mathscr{K}$ can be expressed in the form

$$\left.\begin{aligned} x &= p(2) + q(3) + r(0) \\ y &= p(5) + q(1) + r(3) \end{aligned}\right\} \quad \text{where} \quad \begin{cases} p \geq 0, q \geq 0, r \geq 0 \\ \text{and} \\ p + q + r = 1. \end{cases}$$

This is called a *convex combination* of the coordinates (2, 5), (3, 1), and (0, 3). The set of all convex combinations is the *convex hull* of the given points. Clearly, these remarks are readily generalized so as to apply to m points representing the rows of any $m \times 2$ matrix game.

Returning to our illustration, *player A seeks that point of the strategy polygon $\mathcal{K}$, whose smaller coordinate is as large as possible.* In this case he need merely draw the line l defined by $y = x$, and find that point where it intersects $\mathcal{K}$ furthest from the origin. This is the point T in Figure 7*b*. All other points of $\mathcal{K}$ have either a

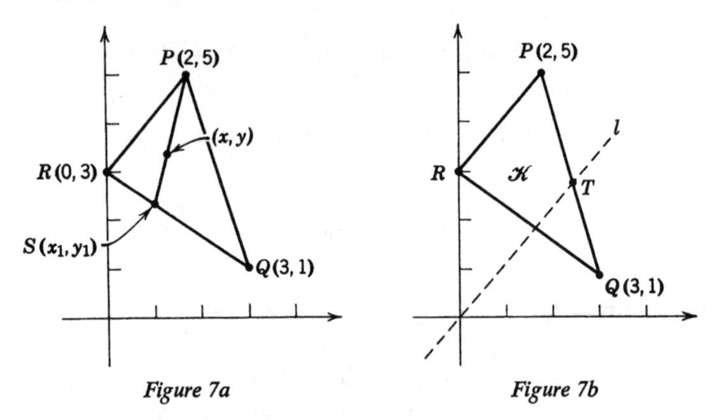

Figure 7a *Figure 7b*

smaller first coordinate or a smaller second coordinate. Using parametric equations for segment $\overset{\bullet\!-\!\bullet}{PQ}$.

$$x = t(2) + (1 - t)(3) = 3 - t$$
$$y = t(5) + (1 - t)(1) = 1 + 4t$$

and equating x and y, we obtain $3 - t = 1 + 4t$

therefore $\qquad t = \frac{2}{5}, \qquad (1 - t) = \frac{3}{5},$

and the common value for x and y becomes $x = y = \frac{13}{5}$. These are the coordinates of point T. Thus it appears that A's optimal strategy is $(\frac{2}{5}, \frac{3}{5}, 0)$ and the value of the game (to A) is $\frac{13}{5}$. This will also be B's minimum expected payoff according to the Minimax Theorem. B's optimal strategy can also be determined from the diagram by a geometrical argument (see Reference 19, pages 14–19). However, we shall not pursue this because we shall presently make available the more powerful techniques of linear programming for the solution of this and even more general problems.

It is by no means accidental that A did not have to use his third pure strategy (0, 0, 1) in the previous example. His optimal strategy was a "mixture" of only two out of his three pure strategies i.e., a mixture of (1, 0, 0) and (0, 1, 0) only. This is due to the fact that the smaller coordinate of the point T is required to be as large as possible and this requirement forces T to lie on the boundary of $\mathscr{K}$. For, if T were in the interior of the polygon, then any line through T with a *positive* slope would contain new points of $\mathscr{K}$ having both

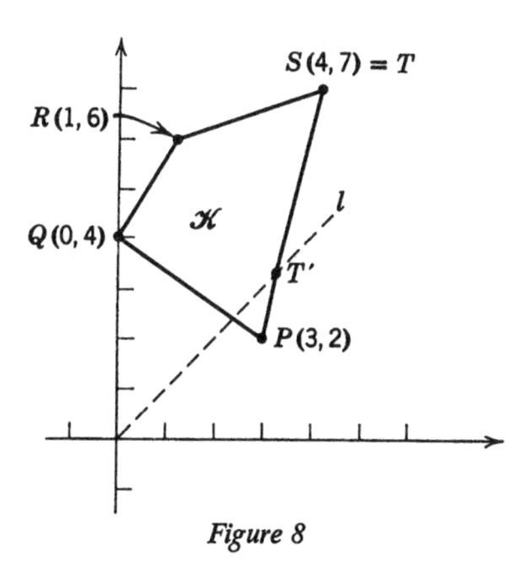

Figure 8

coordinates larger than the corresponding coordinates of T and this would contradict the definition of T. The argument clearly applies to any $m \times 2$ matrix game and shows that *player A need never use a mixture of more than two of his m available pure strategies, so long as B is confined to two pure strategies.* Similarly, in a $2 \times n$ matrix game, player B need never use a mixture of more than two of the n available pure strategies.

In special cases only one pure strategy may, of course, suffice. For example, consider the 4×2 matrix game.

$$\textit{Player A} \quad \overset{\textit{Player B}}{\begin{bmatrix} 3 & 2 \\ 0 & 4 \\ 1 & 6 \\ 4 & 7 \end{bmatrix}}$$

The rows of this matrix are plotted as points P, Q, R, and S in Figure 8. The strategy polygon for player A is a quadrilateral $\mathscr{K}$, with vertices at P, Q, R, and S. From the origin, we proceed along the line l out to its furthest intersection with $\mathscr{K}$. This yields the point T' in the figure. However, even though T' is on the boundary, there are still further points of $\mathscr{K}$ which can be reached from T' along lines of positive slope. In fact, proceeding upward along the edge from T', we reach vertex S which is the desired point of $\mathscr{K}$. Therefore, player A's optimal strategy is the pure strategy (0, 0, 0, 1). The value of the game is the smaller coordinate of S, namely 4, and B's

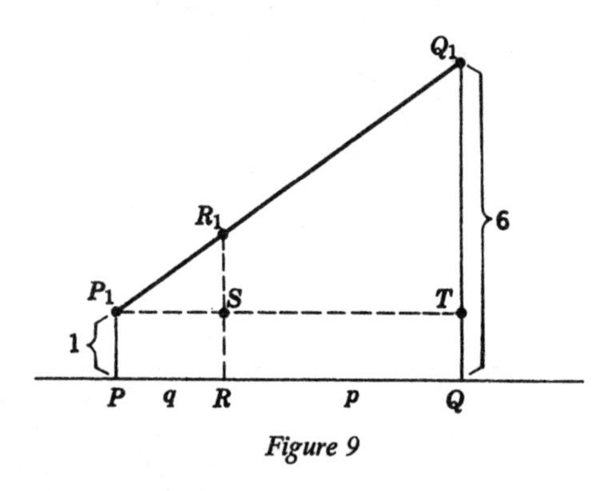

Figure 9

optimal strategy is therefore (1, 0). The existence of optimal pure strategies for both players shows that the game is strictly determined and has in fact a saddle point at row 4, column 1.

We shall conclude this section by describing still another graphical technique which works well for $m \times 2$ or $2 \times n$ matrix games. Consider, for example, the 3×2 matrix game.

$$\text{Player } A \quad \overset{\text{Player } B}{\begin{bmatrix} 1 & 6 \\ 4 & 5 \\ 5 & 3 \end{bmatrix}}$$

Here player B has just two pure strategies and can "mix" them in any proportions. In Figure 9 the two pure strategies (1, 0) and (0, 1) are represented by points P and Q along a horizontal line (the x-axis). The distance from P to Q is conveniently designated as one unit along this line. Any of B's (pure or mixed) strategies (p, q), where

$p \geq 0$, $q \geq 0$, and $p + q = 1$, may be represented by (an intermediate) point R, which divides segment $\overset{\bullet\!-\!\bullet}{PQ}$ in the ratio $q : p$.* As we have chosen $PQ = 1$ unit, we may write

$$PR = q, \qquad RQ = p$$

For example, if B adopts the mixed strategy $(\frac{2}{3}, \frac{1}{3})$ then $PR = \frac{1}{3}$ and $RQ = \frac{2}{3}$, so in this case R is located two thirds of the way from point Q to point P.

Now consider player A's first pure strategy against which the payoffs are 1 and 6. Mark point P_1 directly above point P at a height of 1 unit and mark point Q_1 directly above point Q at a height of 6 units. (The vertical scale need not be the same as the horizontal scale.) Now if R_1 is any point of segment $\overset{\bullet\!-\!\bullet}{P_1Q_1}$, then the height of R_1 will denote the payoff which B should expect to make against A's first strategy, if B adopts the mixed strategy corresponding to point R directly below R_1. This is geometrically evident, using similar triangles in Figure 9:

$$\frac{SR_1}{q} = \frac{TQ_1}{q + p}, \quad \therefore SR_1 = q(TQ_1)$$

$$\begin{aligned} \therefore \quad RR_1 &= RS + SR_1 \\ &= PP_1 + q(TQ_1) \\ &= PP_1 + q(QQ_1 - QT) \\ &= PP_1 + q(QQ_1 - PP_1) \\ &= (1 - q)(PP_1) + q(QQ_1), \end{aligned}$$

i.e.,

$$RR_1 = p(PP_1) + q(QQ_1). \qquad \text{Q.E.D.}$$

In exactly the same manner, let segment $\overset{\bullet\!-\!\bullet}{P_2Q_2}$ be drawn using the second row entries 4 and 5 as heights for its endpoints, and let segment $\overset{\bullet\!-\!\bullet}{P_3Q_3}$ be drawn using the third row payoffs 5 and 3. These segments now yield B's expected payoffs against A's second and third pure strategies at all intermediate points between P and Q (see Figure 10).

* If $q = 0$, R coincides with P. If $p = 0$, R coincides with Q.

Let R be any such intermediate point chosen by B. A will naturally prefer that strategy which yields the greatest vertical height. In this particular game, if B's strategy point R lies to the left of the point S indicated in Figure 10, then clearly A should use his third pure strategy; if B's strategy point R' lies to the right of point T, then A should use his first pure strategy; if B's strategy point R'' lies between points S and T, then A had best use his second pure strategy. An argument exactly like that given on page 98 will convince the reader that in each of these situations A need not

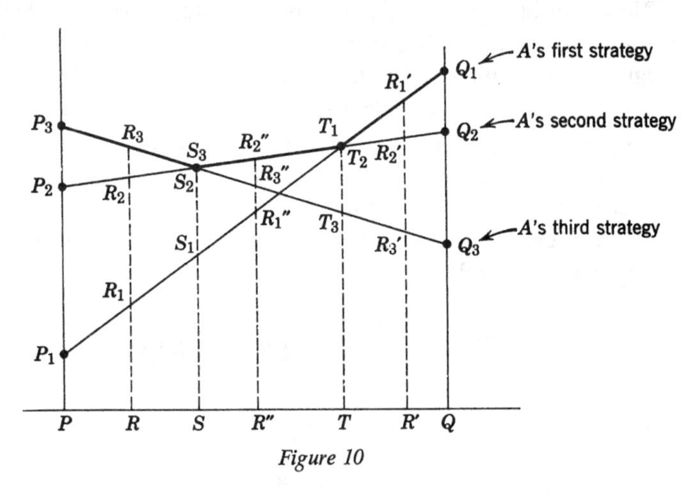

Figure 10

resort to a mixed strategy. However, B will naturally try to keep the greatest vertical height to a minimum. He will clearly achieve this at point S. The reader can now verify that this point corresponds to the mixed strategy $(\frac{2}{3}, \frac{1}{3})$ (see Exercise 5, below). This is actually B's optimal strategy and yields a vertical height of $\frac{13}{3}$ (value of the game to A). A's optimal strategy can be shown to be a mixture of his second and third pure strategies. This is also left as an exercise for the reader (see Exercise 6, below).

A similar analysis ("dual" to this one) applies whenever player A, rather than B, has only two pure strategies. The reader will find it profitable to think this through on his own. We shall not pursue these graphical techniques as it is time to move on to the very fascinating general treatment which linear programming can lend to the Theory of Games.

Exercises

1. Solve each of the following 2×2 matrix games using the first graphical method of this section, i.e., draw the strategy segments for each player, note the relation of each of these segments to the line l defined by $y = x$, and then determine the optimal strategies and the value of the game.

(*a*) $\begin{bmatrix} 1 & 0 \\ 0 & 3 \end{bmatrix}$ (*b*) $\begin{bmatrix} 4 & 1 \\ 1 & 7 \end{bmatrix}$

(*c*) $\begin{bmatrix} 1 & 3 \\ 4 & -1 \end{bmatrix}$ (*d*) $\begin{bmatrix} 2 & 3 \\ 1 & 5 \end{bmatrix}$

(*e*) $\begin{bmatrix} -2 & -4 \\ 3 & 5 \end{bmatrix}$ (*f*) $\begin{bmatrix} 5 & 1 \\ 3 & 4 \end{bmatrix}$

2. Solve each of the following 3×2 matrix games by plotting the strategy polygon $\mathscr{K}$ for player A and determining that point of $\mathscr{K}$ whose smaller coordinate is as large as possible. In each case, determine A's optimal strategy and the value of the game.

(*a*) $\begin{bmatrix} 4 & 2 \\ 1 & 3 \\ 3 & 4 \end{bmatrix}$ (*b*) $\begin{bmatrix} 3 & 2 \\ 6 & 4 \\ 4 & 5 \end{bmatrix}$

(*c*) $\begin{bmatrix} 1 & 5 \\ 3 & 4 \\ 2 & 1 \end{bmatrix}$ (*d*) $\begin{bmatrix} -3 & 6 \\ 6 & 3 \\ 8 & -2 \end{bmatrix}$

3. Solve each of the following 2×3 matrix games by plotting the strategy polygon for player B and determining that point of this polygon whose larger coordinate is as small as possible. (This argument is the "dual" of the previous argument.) In each case, determine B's optimal strategy and the value of the game.

(*a*) $\begin{bmatrix} 3 & 0 & 2 \\ 1 & 2 & 3 \end{bmatrix}$ (*b*) $\begin{bmatrix} 5 & 3 & 2 \\ 3 & 4 & 1 \end{bmatrix}$ (*c*) $\begin{bmatrix} 4 & 3 & 5 \\ 1 & 2 & 3 \end{bmatrix}$

4. Solve each of the games in Exercises 1, 2, and 3, by using the second graphical method described in the text. Note that when player A has only two pure strategies, and these are plotted along the horizontal axis, the reasoning is "dual" to that used in the text.

5. Verify that in Figure 10, $PS = \frac{1}{3}$ and $SS_2 = \frac{13}{3}$. (*Hint:* Choose an origin at point P and express equations for lines $\overleftrightarrow{P_2Q_2}$ and $\overleftrightarrow{P_3Q_3}$ in the form $y = mx + b$; then solve simultaneously.)

6. Calculate A's optimal strategy in the problem represented by Figure 10. (Hint: Use the following argument: Whatever B does, the payoff to A must be the value of the ordinate at S. Hence, draw a parallel to the x-axis through the point S_2 and determine the ratio in which it divides the segment $\overline{P_2P_3}$ (or the segment $\overline{Q_2Q_3}$). Why does this ratio represent A's optimal strategy?)

(*Answers*)

1. (*a*) A: $(\frac{3}{4}, \frac{1}{4})$, B: $(\frac{3}{4}, \frac{1}{4})$, $v = \frac{3}{4}$.
   (*b*) A: $(\frac{2}{3}, \frac{1}{3})$, B: $(\frac{1}{3}, \frac{2}{3})$, $v = 3$.
   (*c*) A: $(\frac{5}{7}, \frac{2}{7})$, B: $(\frac{4}{7}, \frac{3}{7})$, $v = \frac{13}{7}$.
   (*d*) A: $(1, 0)$, B: $(1, 0)$, $v = 2$.
   (*e*) A: $(0, 1)$, B: $(1, 0)$, $v = 3$.
   (*f*) A: $(\frac{1}{5}, \frac{4}{5})$, B: $(\frac{3}{5}, \frac{2}{5})$, $v = \frac{17}{5}$.
2. (*a*) A: $(\frac{1}{2}, 0, \frac{1}{2})$, $v = \frac{9}{2}$.
   (*b*) A: $(0, \frac{1}{3}, \frac{1}{3})$, $v = \frac{14}{3}$.
   (*c*) A: $(0, 1, 0)$, $v = 3$.
   (*d*) A: $(\frac{3}{14}, \frac{11}{14}, 0)$, $v = \frac{57}{14}$.

PART FIVE

Matrix Games and Linear Programming

1. The Solution of Matrix Games by Linear Programming

In Part Four we applied a graphical analysis to the 3×2 matrix game:

$$\textit{Player } A \quad \overset{\textit{Player } B}{\begin{bmatrix} 2 & 5 \\ 3 & 1 \\ 0 & 3 \end{bmatrix}}$$

Let us now re-examine this game from a purely algebraic point of view.

Denote any (pure or mixed) strategy for A by (p, q, r), so that

$$p \geq 0, \quad q \geq 0, \quad r \geq 0, \quad \text{and} \tag{1}$$

$$p + q + r = 1. \tag{2}$$

A's expected payoffs against each of B's two *pure* strategies are respectively given by

$$2p + 3q + 0r \quad \text{and} \quad 5p + 1q + 3r.$$

Now, for each (p, q, r), let g denote the *smaller* of these two payoffs (or their common value if they are the same). Then certainly

$$\begin{cases} 2p + 3q + 0r \geq g \\ 5p + 1q + 3r \geq g. \end{cases} \tag{3}$$

In this particular problem, g is obviously positive because all the entries in the payoff matrix are ≥ 0. In general, if any of these entries are negative, we can make them *all* ≥ 0 by simply adding a suitable constant k to all of the entries. (Take k = the absolute value of the smallest entry.) For each particular strategy, including the sought for optimal ones, this would merely increase the expected payoff by k. It would not, therefore, alter the optimal strategies. In short, there is no loss of generality in assuming g is positive.

Returning to our problem, we divide (1), (2), and (3) throughout by g and obtain

$$\left(\frac{p}{g}\right) \geq 0, \quad \left(\frac{q}{g}\right) \geq 0, \quad \left(\frac{r}{g}\right) \geq 0. \tag{1'}$$

$$\left(\frac{p}{g}\right) + \left(\frac{q}{g}\right) + \left(\frac{r}{g}\right) = \frac{1}{g}. \tag{2'}$$

$$\begin{cases} 2\left(\frac{p}{g}\right) + 3\left(\frac{q}{g}\right) + 0\left(\frac{r}{g}\right) \geq 1 \\ 5\left(\frac{p}{g}\right) + 1\left(\frac{q}{g}\right) + 3\left(\frac{r}{g}\right) \geq 1. \end{cases} \tag{3'}$$

A wishes to choose his strategy (p, q, r) so as to *maximize* g. Clearly, this can be done by *minimizing* $(1/g)$. For simplicity, let us change the notation as follows: Let

$$(4) \qquad x = \frac{p}{g}, \quad y = \frac{q}{g}, \quad z = \frac{r}{g}, \quad m = \frac{1}{g}.$$

Then A wishes to

$$(5) \qquad \text{determine:} \quad x \geq 0, \quad y \geq 0, \quad z \geq 0$$

$$(6) \qquad \text{so that:} \quad \begin{cases} 2x + 3y + 0z \geq 1 \\ 5x + 1y + 3z \geq 1 \end{cases}$$

$$(7) \qquad \text{and so that:} \quad x + y + z = m \quad \text{is a MINIMUM.}$$

A's problem therefore reduces to a linear programming problem which can readily be solved by the Simplex Method of Chapter 3!

Before doing this, however, let us look at B's problem, again adopting a purely algebraic point of view. Denote any strategy for B by (s, t), so that

$$(8) \qquad s \geq 0, \quad t \geq 0, \quad \text{and}$$

$$(9) \qquad s + t = 1.$$

Against each of A's three *pure* strategies, B must expect to pay off the amounts respectively given by

$$2s + 5t, \quad 3s + 1t, \quad 0s + 3t.$$

Now, for each (s, t), let h denote the *largest* of these three payoffs (or the common value if two or more are "tie"). Then:

$$(10) \qquad \begin{cases} 2s + 5t \leq h \\ 3s + 1t \leq h \\ 0s + 3t \leq h. \end{cases}$$

In this particular problem, h is also positive (for the same reasons as were mentioned for g) and may be assumed positive, without loss of generality, in all problems. We may therefore divide (8), (9), and (10) throughout by h to obtain

$$(8') \qquad \left(\frac{s}{h}\right) \geq 0, \quad \left(\frac{t}{h}\right) \geq 0, \quad \text{and}$$

$$(9') \qquad \left(\frac{s}{h}\right) + \left(\frac{t}{h}\right) = \frac{1}{h}$$

$$(10') \qquad \begin{cases} 2\left(\frac{s}{h}\right) + 5\left(\frac{t}{h}\right) \leq 1 \\ 3\left(\frac{s}{h}\right) + 1\left(\frac{t}{h}\right) \leq 1 \\ 0\left(\frac{s}{h}\right) + 3\left(\frac{t}{h}\right) \leq 1. \end{cases}$$

Player B wishes to choose his strategy (s, t) so as to *minimize* h. Clearly this can be accomplished by *maximizing* $1/h$.

Once again, for simplicity, we change the notation. Let

$$(11) \qquad u = \frac{s}{h}, \quad v = \frac{t}{h}, \quad M = \frac{1}{h}.$$

Then B wishes to

(12) determine: $u \geq 0, \quad v \geq 0$

$$(13) \qquad \text{so that:} \quad \begin{cases} 2u + 5v \leq 1 \\ 3u + 1v \leq 1 \\ 0u + 3v \leq 1 \end{cases}$$

(14) and so that: $u + v = M$ is a MAXIMUM.

Consequently, *B's problem is also a linear programming problem,* which, in the present case, is even simpler than A's. It can, in fact, be solved using the simple graphical method of Chapter 1. However, it will be more illuminating to solve both this problem as well as A's problem by the Simplex Method.

Tackling B's problem first, we introduce three slack variables a, b, and c into (13) so as to convert these inequalities to equations. We also multiply both sides of (14) by (-1) and bring all terms to the left member. We obtain

$$(12') \qquad u \geq 0, \quad v \geq 0, \quad w \geq 0, \quad a \geq 0, \quad b \geq 0, \quad c \geq 0.$$

$$(13') \qquad \begin{aligned} 2u + 5v + a \qquad\qquad &= 1 \\ 3u + 1v \qquad + b \qquad &= 1 \\ 0u + 3v \qquad\qquad + c &= 1 \end{aligned}$$

$$(14') \qquad -u - v \qquad\qquad + M = 0$$

the object being to maximize M. The sequence of tableaux is

	u	v	a	b	c	M			Comments
$\frac{1}{2}=\frac{1}{2}$	2	5	1	0	0	0	1	$=a$	
$\rightarrow \frac{1}{3}=\frac{1}{3}$	(3)	1	0	1	0	0	1	$=b$	Although there is a "tie" for most negative entry in the last row, we ignore this tie and arbitrarily select the variable u to enter the basis.
	0	3	0	0	1	0	1	$=c$	
	-1	-1	0	0	0	1	0	$=M$	
	↑								

$\rightarrow \frac{1}{3}/\frac{13}{3}=\frac{1}{13}$	0	$(\frac{13}{3})$	1	$-\frac{2}{3}$	0	0	$\frac{1}{3}$	$=a$	u is now in the basis but the presence of the negative entry $-\frac{2}{3}$ in the last row indicates that M is not yet a maximum. Accordingly, we next introduce v into the basis.
$\frac{1}{3}/\frac{1}{3}=1$	1	$\frac{1}{3}$	0	$\frac{1}{3}$	0	0	$\frac{1}{3}$	$=u$	
$\frac{1}{3}=\frac{1}{3}$	0	3	0	0	1	0	1	$=c$	
	0	$-\frac{2}{3}$	0	$\frac{1}{3}$	0	1	$\frac{1}{3}$	$=M$	
		↑							

0	1	$\frac{3}{13}$	$-\frac{2}{13}$	0	0	$\frac{1}{13}$	$=v$	
1	0	$-\frac{1}{13}$	$\frac{5}{13}$	0	0	$\frac{4}{13}$	$=u$	Final tableau. (no more negative entries in last row).
0	0	$-\frac{9}{13}$	$\frac{6}{13}$	1	0	$\frac{10}{13}$	$=c$	
0	0	$\frac{2}{13}$	$\frac{3}{13}$	0	1	$\frac{5}{13}$	$=M$	

The final tableau yields the following optimum solution for player B:

$$M=\tfrac{5}{13};\quad \therefore\ h=1/M=\tfrac{13}{5}$$

$$u=\tfrac{4}{13};\quad \therefore\ s=hu\ =\tfrac{4}{5}$$

$$v=\tfrac{1}{13};\quad \therefore\ t=hv\ =\tfrac{1}{5}.$$

This means that B's optimal strategy is $(\frac{4}{5}, \frac{1}{5})$, the associated expected payoff to A, being $\frac{13}{5}$.

Recall that by means of the graphical analysis of the preceding section we obtained $\frac{13}{5}$ as the value of the game *to player A* (see above, page 104). We have now succeeded in proving that this same amount, $\frac{13}{5}$, also represents the value of the game *to player B*. We have therefore verified the Minimax Theorem for this particular problem. Moreover, we have calculated *B*'s optimal strategy and thus completed the solution of this game.

Our present method looks particularly elegant using condensed tableaux:

Player B

u	v		
2	5	1	a
3	1	1	b
0	3	1	c
-1	-1	0	M

(Initial Condensed Tableau)

(First Pivotal Exchange: u and b)

b	v		
$-\frac{2}{3}$	$\frac{13}{3}$	$\frac{1}{3}$	a
$\frac{1}{3}$	$\frac{1}{3}$	$\frac{1}{3}$	u
0	3	1	c
$\frac{1}{3}$	$-\frac{2}{3}$	$\frac{1}{3}$	M

(Second Pivotal Exchange: v and a)

b	a		
$-\frac{2}{13}$	$\frac{3}{13}$	$\frac{1}{13}$	v
$\frac{5}{13}$	$-\frac{1}{13}$	$\frac{4}{13}$	u
$\frac{6}{13}$	$-\frac{9}{13}$	$\frac{10}{13}$	c
$\frac{3}{13}$	$\frac{2}{13}$	$\frac{5}{13}$	M

Final (Optimal) Tableau

Observe that *the initial tableau consists of the original payoff matrix, augmented by an additional column of 1's, an additional row of -1's and an additional 0 in the lower right hand corner.* A sequence of pivotal exchanges applied to this augmented matrix yields the solution to *both A*'s and *B*'s problem, as we shall presently see.

2. Duality and the Minimax Theorem

Let us return now to A's problem and solve it by the Simplex Method. In Equation (7) we let $m = -M'$ and bring all terms to the left side, thus

$$x + y + z + M' = 0.$$

In order to minimize m, we must *maximize* M'. We introduce two non-negative slack variables d and e into (6) so as to convert these inequalities into equations. Then we append two non-negative artificial variables k and l, one to each of these equations, to supply us with a starting basis. The artificial variables are assigned an arbitrarily large positive coefficient P in the last equation. We now have the following (artificial) problem:

$$(5') \qquad x \geq 0, \quad y \geq 0, \quad z \geq 0, \quad d \geq 0, \quad e \geq 0, \quad k \geq 0, \quad l \geq 0$$

$$(6') \qquad \begin{cases} 2x + 3y + 0z - d \qquad + k \qquad = 1 \\ 5x + 1y + 3z \qquad - e \qquad + l \qquad = 1 \end{cases}$$

$$(7') \qquad x + y + z \qquad + Pk + Pl + M' = 0$$

the object being to *maximize* M'.

The starting tableau looks like this.

x	y	z	d	e	k	l	M'	
2	3	0	−1	0	1	0	0	1
5	1	3	0	−1	0	1	0	1
1	1	1	0	0	0	0	1	0
					P	P		

There is no obvious B.F.S. To obtain one, we multiply the first two rows by $-P$ and add both of these results to the last row (i.e., to the "double row"). The resulting tableau does contain a B.F.S. involving the artificial variables and serves as the starting point for our subsequent iterations.

	x	y	z	d	e	k	l	M'			
$\frac{1}{2}$	2	3	0	-1	0	1	0	0	1	$= k$	
$\rightarrow \frac{1}{5}$	(5)	1	3	0	-1	0	1	0	1	$= l$	(Initial B.F.S.)
	1	1	1	0	0	0	0	1	0	$= M'$	
	$-7P$	$-4P$	$-3P$	$1P$	$1P$				$-2P$		
	↑										

	x	y	z	d	e	k	l	M'			
$\rightarrow \frac{3}{13}$	0	($\frac{13}{5}$)	$-\frac{6}{5}$	-1	$\frac{2}{5}$	1	$-\frac{2}{5}$	0	$\frac{3}{5}$	$= k$	(x enters the basis displacing l.)
1	1	$\frac{1}{5}$	$\frac{3}{5}$	0	$-\frac{1}{5}$	0	$\frac{1}{5}$	0	$\frac{1}{5}$	$= x$	
	0	$\frac{4}{5}$	$\frac{2}{5}$	0	$\frac{1}{5}$	0	$-\frac{1}{5}$	1	$-\frac{1}{5}$	$= M'$	
		$-\frac{13}{5}P$	$\frac{6}{5}P$	$1P$	$-\frac{2}{5}P$		$\frac{3}{5}P$		$-\frac{3}{5}P$		
		↑									

x	y	z	d	e	k	l	M'			
0	1	$-\frac{6}{13}$	$-\frac{5}{13}$	$\frac{2}{13}$	$\frac{5}{13}$	$-\frac{2}{13}$	0	$\frac{3}{13}$	$= y$	(y enters the basis displacing k.)
1	0	$\frac{9}{13}$	$\frac{1}{13}$	$-\frac{3}{13}$	$-\frac{1}{13}$	$\frac{3}{13}$	0	$\frac{2}{13}$	$= x$	
0	0	$\frac{10}{13}$	$\frac{4}{13}$	$\frac{1}{13}$	$-\frac{4}{13}$	$-\frac{2}{13}$	1	$-\frac{5}{13}$	$= M'$	
					$1P$	$1P$				

The last tableau yields a B.F.S. in terms of the *original* variables x and y. The artificial variables k and l, are no longer in the basis and may now be dropped. The original variable z is retained with a value of 0, since it is not in the basis. The slack variables d and e are also nonbasic. In a more complicated problem we would use this B.F.S. as the starting point for further iterations (Phase II) using only the original and slack variables. However in the present case this is unnecessary, as we already have an optimal tableau. In fact the optimal solution to A's problem is given by

$$m = -M' = \tfrac{5}{13}, \quad \therefore g = 1/m = \tfrac{13}{5}.$$
$$x = \tfrac{2}{13}, \quad \therefore p = gx = \tfrac{2}{5}.$$
$$y = \tfrac{3}{13}, \quad \therefore q = gy = \tfrac{3}{5}.$$
$$z = 0, \quad \therefore r = gz = 0.$$

A's optimal strategy is therefore $(\frac{2}{5}, \frac{3}{5}, 0)$, and the associated payoff is $\frac{13}{5}$. This is all in complete agreement with the results already obtained earlier by graphical analysis in Part Four, Section 3.

We could, of course, solve A's problem using condensed tableaux.

Although not as elegant as *B*'s, these tableaux are of interest.

Player A

(Initial Condensed Tableau)

	x	y	z	d	e		
$\frac{1}{2}$	2	3	0	-1	0	1	k
$\rightarrow \frac{1}{5}$	(5)	1	3	0	-1	1	l
	1	1	1	0	0	0	M'
	$-7P$	$-4P$	$-3P$	$1P$	$1P$	$-2P$	
	↑						

(First Pivotal Exchange: l and x)

	l	y	z	d	e		
$\rightarrow \frac{3}{13}$	$-\frac{2}{5}$	$(\frac{13}{5})$	$-\frac{6}{5}$	-1	$\frac{2}{5}$	$\frac{3}{5}$	k
1	$\frac{1}{5}$	$\frac{1}{5}$	$\frac{3}{5}$	0	$-\frac{1}{5}$	$\frac{1}{5}$	x
	$-\frac{1}{5}$	$\frac{4}{5}$	$\frac{2}{5}$	0	$\frac{1}{5}$	$-\frac{1}{5}$	M'
	$\frac{7}{5}P$	$-\frac{13}{5}P$	$\frac{6}{5}P$	$1P$	$-\frac{2}{5}P$	$-\frac{3}{5}P$	
		↑					

(Second Pivotal Exchange: k and y)

Final (Optimal) Tableau

l	k	z	d	e		
$-\frac{2}{13}$	$\frac{5}{13}$	$-\frac{6}{13}$	$-\frac{5}{13}$	$\frac{2}{13}$	$\frac{3}{13}$	y
$\frac{3}{13}$	$-\frac{1}{13}$	$\frac{9}{13}$	$\frac{1}{13}$	$-\frac{3}{13}$	$\frac{2}{13}$	x
$-\frac{1}{13}$	$-\frac{4}{13}$	$\frac{10}{13}$	$\frac{4}{13}$	$\frac{1}{13}$	$-\frac{5}{13}$	M'
$1P$	$1P$					

The final (optimal) tableau for *A*'s problem is the portion of the matrix to the right of the dotted line (i.e., discarding the artificial variables *k* and *l*). This tableau bears a remarkable relation to the final (optimal) tableau for *B*'s problem. Let us array these two tableaux side by side:

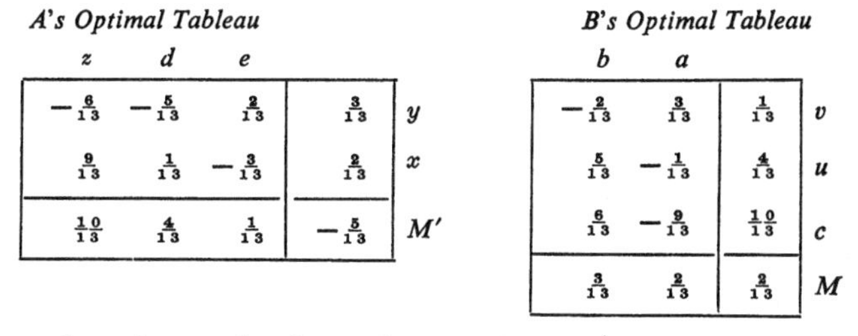

A's Optimal Tableau

z	d	e		
$-\frac{6}{13}$	$-\frac{5}{13}$	$\frac{2}{13}$	$\frac{3}{13}$	y
$\frac{9}{13}$	$\frac{1}{13}$	$-\frac{3}{13}$	$\frac{2}{13}$	x
$\frac{10}{13}$	$\frac{4}{13}$	$\frac{1}{13}$	$-\frac{5}{13}$	M'

B's Optimal Tableau

b	a		
$-\frac{2}{13}$	$\frac{3}{13}$	$\frac{1}{13}$	v
$\frac{5}{13}$	$-\frac{1}{13}$	$\frac{4}{13}$	u
$\frac{6}{13}$	$-\frac{9}{13}$	$\frac{10}{13}$	c
$\frac{3}{13}$	$\frac{2}{13}$	$\frac{2}{13}$	M

Observe that the entries in each *row* of *A*'s tableau are identical, except for certain changes of sign, with the entries in a suitably chosen *column* of *B*'s tableau. By rearranging the rows and columns a bit further, we can see a more definite pattern emerge.

(15)

A

d	e	z		
$\frac{1}{13}$	$-\frac{3}{13}$	$\frac{9}{13}$	$\frac{2}{13}$	x
$-\frac{5}{13}$	$\frac{2}{13}$	$-\frac{6}{13}$	$\frac{3}{13}$	y
$\frac{4}{13}$	$\frac{1}{13}$	$\frac{10}{13}$	$-\frac{5}{13}$	M'

B

a	b		
$-\frac{1}{13}$	$\frac{5}{13}$	$\frac{4}{13}$	u
$\frac{3}{13}$	$-\frac{2}{13}$	$\frac{1}{13}$	v
$-\frac{9}{13}$	$\frac{6}{13}$	$\frac{10}{13}$	c
$\frac{2}{13}$	$\frac{3}{13}$	$\frac{5}{13}$	M

*The matrices in the upper left and lower right hand corners of either tableau are "negative transposes" of the corresponding matrices of the other tableau. The matrices in lower left and upper right corners are simply transposes of their correspondents.**

This pattern is not accidental. It exhibits a fundamental phenomenon known as "*duality*," which provides new insights into the theory of the Simplex Method.

Remember that the tableaux in (15) are *condensed.* If we refer back to *B*'s *extended* optimal tableau (see page 115 above), we can translate each row into an equation.

$$
(16)\qquad
\begin{array}{r|l}
u & -\tfrac{1}{13}a + \tfrac{5}{13}b = \tfrac{4}{13} \\
v & +\tfrac{3}{13}a - \tfrac{2}{13}b = \tfrac{1}{13} \\
c & -\tfrac{9}{13}a + \tfrac{6}{13}b = \tfrac{10}{13} \\
M & +\tfrac{2}{13}a + \tfrac{3}{13}b = \tfrac{5}{13}
\end{array}
\qquad (B\text{'s Equations})
$$

Similarly, if we refer back to *A*'s extended optimal tableau (see page 118 above), then dropping the artificial variables *k* and *l*, we can likewise translate each row into an equation.

$$
(17)\qquad
\begin{array}{r|l}
x & +\tfrac{1}{13}d - \tfrac{3}{13}e + \tfrac{9}{13}z = \tfrac{2}{13} \\
y & -\tfrac{5}{13}d + \tfrac{2}{13}e - \tfrac{6}{13}z = \tfrac{3}{13} \\
M' & +\tfrac{4}{13}d + \tfrac{1}{13}e + \tfrac{10}{13}z = -\tfrac{5}{13}
\end{array}
\qquad (A\text{'s Equations})
$$

* The *transpose* of a matrix is the matrix formed by changing its rows to columns and its columns to rows. The *negative* of a matrix is formed by changing the signs of all entries.

The *duality pattern* which we observed in the condensed tableaux in (15) is evidently a property of the coefficients enclosed within the dotted lines in the two systems of Equations (16) and (17). We shall call two such systems *dual systems.*

If the occurrence of the duality pattern was not accidental, where did it come from? The explanation lies in the fact that the two sets of equations (16) and (17) were derived from, and are respectively *equivalent* to, the two original systems of equations, namely (13′), (14′) in the case of B's problem and (6′), (7′) in the case of A's problem, the artificial variables being taken as 0 in the latter case. The equivalences stem from the fact that each iteration, i.e., each transition from one tableau to the next, is actually a Gauss-Jordan elimination process. *This process always transforms a system of linear equations into an equivalent system!* Now the two original systems can be rewritten as follows:

(18)

	B's equations		*A's equations*
a	$+2u + 5v = 1$	d	$-2x - 3y + 0z = -1$
b	$+3u + 1v = 1$	e	$-5x - 1y - 3z = -1$
c	$+0u + 3v = 1$	M'	$+1x + 1y + 1z = 0$
M	$-1u - 1v = 0$		

It is apparent that the duality pattern was a property of the original equations from the very start! The algebraic formulation of A's and B's problem in a matrix game took the form of a pair of dual systems of linear equations. Each of these systems has been transformed by the Simplex Method into an equivalent system, and the resulting systems are again dual systems.

A detailed analysis of duality phenomena is beyond the scope of this monograph.* However, we shall sketch briefly, some of the ideas involved. Given a dual pair of systems, such as (18) above, a pivot is chosen in one of these systems (for example the coefficient of u in the second of B's equations). The basic variable in the row of the pivot (this is b in our example) must exchange places with the nonbasic variable in the column of the pivot (this is u in our example). This is neatly accomplished by the pivotal exchange method using condensed tableaux. Now if one selects the corresponding pivot in

* See, for example, Reference 19 (Vajda) Chapter VI, pp. 46–52 and Chapter VIII, pp. 72–74.

the condensed tableau for the dual system, and performs the corresponding pivotal exchange (this exchanges d and y in the present example), the duality pattern will be maintained, i.e., *corresponding pivotal exchanges transform dual systems into dual systems!*

We can prove this important assertion fairly easily and obtain, incidentally, another remarkable result as well. Consider any pair of dual systems, such as (18) above, and let us refer to one of these systems as the "primal," the other as its "dual." (For example, in (18) we may consider B's equations as primal and A's equations as dual.) The condensed tableaux for these two systems may be indicated as follows:

(19)

Primal

	⋮		⋮		⋮
...	p (circled)	...	q	...	x
	⋮		⋮		⋮
...	r	...	s	...	y
	⋮		⋮		⋮
...	u	...	v	...	M

Dual

	⋮		⋮		⋮
...	$-p$ (circled)	...	$-r$	...	u
	⋮		⋮		⋮
...	$-q$	...	$-s$	...	v
	⋮		⋮		⋮
...	x	...	y	...	M'

(Temporarily, we do not assume that $M' = -M$. We shall therefore obtain a slightly more general result, of which our assertion will be a special case.) We now transform each of these tableaux using p as a pivot for the primal and $-p$ as a pivot for the dual. In both cases we simply apply the two rules derived in Chapter III, Section 5 (Page 77):

Transformed Primal

	⋮		⋮		⋮
...	$1/p$	...	q/p	...	x/p
	⋮		⋮		⋮
...	$-r/p$	...	$s + q(-r/p)$	...	$y + x(-r/p)$
	⋮		⋮		⋮
...	$-u/p$	...	$v + q(-u/p)$	...	$M + x(-u/p)$

Transformed Dual

	⋮		⋮		⋮
...	$-1/p$	...	r/p	...	$-u/p$
	⋮		⋮		⋮
...	$-q/p$	...	$-s + r(q/p)$	...	$v + u(-q/p)$
	⋮		⋮		⋮
...	x/p	...	$y + r(-x/p)$	...	$M' + u(x/p)$

Not only is the duality pattern immediately evident in the transformed tableaux, but we have the further important result that *$M + M'$ remains constant after each iteration!* In particular, if $M = -M'$ before the iteration, this will still be true after the iteration. In matrix game problems, as well as in many linear programming problems, the initial values assumed by M and M' are both zero. This means that at *every* stage, $M = -M' = m$.

Since this is true at every stage, suppose in particular, that an *optimal* tableau has been achieved for the primal problem, and that this tableau is displayed in (19) above. Because it is an optimal tableau, its last row entries u, v, etc. are all non-negative and the last column entries x, y, etc. are the values of the basic variables, which are therefore also non-negative. These values, in turn, yield the maximum value of M which appears in the lower right corner. Let us turn now to the dual tableau. This is again an optimal tableau, because the last row entries are now x, y, etc. which are all non-negative. The last column entries are now u, v, etc. We therefore have a non-negative set of values for the variables in the dual problem, which yield the maximum value assumed by M'. This value appears in the lower right corner and is identical with the value of $-M$ in the primal optimal tableau. But the maximum value assumed by M' is also the *minimum* value assumed by $-M'$, i.e., by m. We have therefore succeeded in proving that THE MAXIMUM VALUE OF M IN THE PRIMAL PROBLEM IS THE SAME AS THE MINIMUM VALUE OF m IN THE DUAL PROBLEM. Furthermore, we have shown that the solution to both the primal and dual problems can be read off from either of the two dual tableaux. This result is known as the *Fundamental Duality Theorem of Linear Programming*.

The applications of the Fundamental Duality Theorem are far reaching. The Minimax Theorem in the Theory of Games is an obvious corollary. The solution of a matrix game to determine optimal strategies for both players A and B as well as the value of the game, is all accomplished in one optimal condensed tableau, as we have already indicated on page 116. The beauty and power of the Simplex Method are now apparent!

Originating as a convenient computational technique, the Simplex Method has emerged as a fruitful theoretical tool. Moreover, it has the added virtue of being a *constructive* method of proof. For example, in addition to demonstrating the *existence* of optimal solutions in linear programming problems, it provides a practical technique for actually obtaining them. As Dantzig has pointed out, its elegance and essential simplicity stem from the fact that it is completely algebraic, in contrast with the early methods used by von Neumann and others, which required "involved use of topology, functional calculus, and fixed point theorems of L. E. J. Brouwer."*

It is appropriate that we bring our monograph to a close on this note, emphasizing the essentially elementary character of the basic ideas of linear programming. Nevertheless, a deeper study of these ideas must inevitably lead the student into more sophisticated, more advanced areas of mathematics. It is our hope that we have stimulated the reader's interest sufficiently to spur him on towards such further study.

Exercises

1. Players A and B each choose a number 1, 2, or 3, independently of each other. If both choose the same number, then A pays B that amount. If they choose different numbers, then B pays A an amount equal to the number A has chosen.

(*a*) Construct a payoff matrix for this problem.

(*b*) Express each player's problem as a linear programming problem.

(*c*) Set up the dual linear systems which correspond to these two problems.

(*d*) Using condensed tableaux, calculate optimal strategies for both A and B and determine the value of the game.

* George B. Dantzig, "Constructive Proof of the Min-Max Theorem," *Pacific Journal of Mathematics*, Vol. 6, No. 1, 1956, page 25.

Answers: (*a*) $\begin{bmatrix} -1 & 1 & 1 \\ 2 & -2 & 2 \\ 3 & 3 & -3 \end{bmatrix}$ (*d*) A: $(\frac{6}{11}, \frac{3}{11}, \frac{2}{11})$
B: $(\frac{5}{22}, \frac{8}{22}, \frac{9}{22})$
value $= \frac{6}{11}$

2. Consider the following two linear programming problems:

Problem A

Determine: $x \geq 0, \quad y \geq 0, \quad z \geq 0$

so that: $\begin{cases} a_1x + b_1y + c_1z \leq d \\ a_2x + b_2y + c_2z \leq e \end{cases}$

and so that: $ax + by + cy = M$ is a MAXIMUM.

Problem B

Determine: $u \geq 0, \quad v \geq 0$

so that: $\begin{cases} a_1u + a_2v \geq a \\ b_1u + b_2v \geq b \\ c_1u + c_2v \geq c \end{cases}$

and so that: $du + ev = m$ is a MINIMUM.

(*a*) Introduce slack variables and show that the resulting sets of equations are dual systems.

(*b*) Assume that Problem *A* has a finite optimal (feasible) solution. Let the optimal values of the variables be x_0, y_0, z_0 and let $M_0 = ax_0 + by_0 + cz_0$, be the associated (maximum) value of M. Prove that $M_0 \leq m$, *for all feasible values of u and v.*

Hint: Multiply the constraints of Problem *B*, by x_0, y_0, z_0, respectively, add the resulting inequalities, and make use of the constraints in Problem *A*.

(*c*) Assume that Problem *B*, has a finite optimal (feasible) solution. Let the optimal values of the variables u and v be u_0, v_0 and let $m_0 = du_0 + ev_0$ be the associated (minimum) value of m. Prove that $M \leq m_0$, *for all feasible values of x, y, and z.*

(*d*) Observe that (*c*) may be rephrased as follows: *If Problem B has a finite optimal solution, then the values assumed by M in Problem A have an upper bound* (namely m_0). Therefore, let M_0 denote the *least upper bound* (See Part II, Section 4, p. 38) of all the values assumed by M. If there are feasible values of the variables for which *M actually assumes* its least upper bound M (*this can be proved to be true*), then these feasible values constitute an optimal solution for Problem *A*. In short, this proves *one part* of the *Fundamental Duality Theorem* which may be stated as follows:

If either Problem A or Problem B has a finite optimal solution, then the other problem has a finite optimal solution. Moreover the maximum value of M equals the minimum value of m, i.e., $M_0 = m_0$.

(Investigate this problem, by consulting References 2, 8, or 19, if necessary, to supply missing details.)

References

1. Baumol, W. J., *Operations Analysis and Economic Theory*, Prentice-Hall, 1961.
2. Charnes, A., and W. W. Cooper, *Management Models and Industrial Applications of Linear Programming*, John Wiley and Sons, New York, 1961.*
3. Charnes, A., W. W. Cooper, and A. Henderson, *An introduction to Linear Programming*, John Wiley and Sons, New York, 1953.
4. Dorfman, R., P. A. Samuelson, and R. Solow, *Linear Programming and Economic Analysis*, McGraw-Hill Book Co., New York, 1958.
5. Ferguson R., and L. Sargent, *Linear Programming, Fundamentals and Applications*, McGraw-Hill Book Co., New York, 1958.
6. Gale, David, *The Theory of Linear Economic Models*, McGraw-Hill Book Co., New York, 1960.
7. Garvin, G., *Introduction to Linear Programming*, McGraw-Hill Book Co., New York, 1960.
8. Gass, Saul, *Linear Programming, Methods and Applications*, McGraw-Hill Book Co., New York, 1958.
9. Hadley, G., *Linear Programming*, Addison-Wesley Publishing Co., Reading, Mass., 1962.
10. Karlin, S., *Mathematical Methods and Theory in Games, Programming and Economics*, Addison-Wesley Publishing Co., Reading, Mass., 1959.
11. Kemeny, J., J. Snell, and G. Thompson, *Introduction to Finite Mathematics*, Prentice-Hall, Englewood Cliffs, New Jersey, 1957.
12. Kemeny, J., H. Mirkil, J. Snell, and G. Thompson, *Finite Mathematical Structures*, Prentice-Hall, Englewood Cliffs, New Jersey, 1959.
13. Kuhn, H. W., and A. W. Tucker, *Linear Inequalities and Related Systems*, Annals of Mathematics Studies No. 38, Princeton University Press, Princeton, N.J., 1956.
14. McDonald, J., *Strategy in Poker, Business and War*, W. W. Norton and Co., 1951.

* This reference includes an exceptionally extensive bibliography on the subjects of Linear Programming and Theory of Games.

15. Richardson, M., *Fundamentals of Mathematics*, The Macmillan Co., New York, 1958.
16. Rubinshtein, G., *On the Development and Application of Linear Programming in the U.S.S.R.*, Translated by W. Marlowe, and M. Richardson, George Washington University Logistics Research Project, Office of Naval Research, Washington, D.C., 1960.
17. Vajda, S., *Introduction to Linear Programming and the Theory of Games*, John Wiley and Sons, New York, 1960.
18. Vajda, S., *Readings in Linear Programming*, John Wiley and Sons, New York, 1958.
19. Vajda, S., *The Theory of Games and Linear Programming*, John Wiley and Sons, New York, 1956.

Index

A CATALOG OF SELECTED

DOVER BOOKS

IN SCIENCE AND MATHEMATICS

Astronomy

BURNHAM'S CELESTIAL HANDBOOK, Robert Burnham, Jr. Exhaustive guide to the stars beyond our solar system. Andromeda to Cetus in Vol. 1; Chamaeleon to Orion in Vol. 2; and Pavo to Vulpecula in Vol. 3. Hundreds of illustrations. Index in Vol. 3. 2,000pp. 6⅛ x 9¼. 23567-X, 23568-8, 23673-0 Pa., Three-vol. set $46.85

THE EXTRATERRESTRIAL LIFE DEBATE, 1750–1900, Michael J. Crowe. First detailed, scholarly study in English of early ideas about existence of intelligent extraterrestrial life. Examines ideas of Kant, Herschel, Voltaire, Percival Lowell, many others. 16 illustrations. 704pp. 5⅜ x 8½. 40675-X Pa. $19.95

A HISTORY OF ASTRONOMY, A. Pannekoek. Well-balanced study covers Ptolemaic theory, work of Copernicus, Kepler, Newton, Eddington's work on stars, much more. Illustrated. References. 521pp. 5⅜ x 8½. 65994-1 Pa. $15.95

AMATEUR ASTRONOMER'S HANDBOOK, J. B. Sidgwick. Timeless, comprehensive coverage of telescopes, mirrors, lenses, mountings, telescope drives, micrometers, spectroscopes, more. 189 illustrations. 576pp. 5⅝ x 8¼. (Available in U.S. only.) 24034-7 Pa. $13.95

STARS AND RELATIVITY, Ya. B. Zel'dovich and I. D. Novikov. Vol. 1 of *Relativistic Astrophysics* by famed Russian scientists. General relativity, properties of matter under astrophysical conditions, stars and stellar systems. Deep physical insights, clear presentation. 1971 edition. 544pp. 5⅜ x 8½. 69424-0 Pa. $14.95

Chemistry

CHEMICAL MAGIC, Leonard A. Ford. Second Edition, Revised by E. Winston Grundmeier. Over 100 unusual stunts demonstrating cold fire, dust explosions, much more. Text explains scientific principles and stresses safety precautions. 128pp. 5⅜ x 8½. 67628-5 Pa. $5.95

THE DEVELOPMENT OF MODERN CHEMISTRY, Aaron J. Ihde. Authoritative history of chemistry from ancient Greek theory to 20th-century innovation. Covers major chemists and their discoveries. 209 illustrations. 14 tables. Bibliographies. Indices. Appendices. 851pp. 5⅜ x 8½. 64235-6 Pa. $24.95

CATALYSIS IN CHEMISTRY AND ENZYMOLOGY, William P. Jencks. Exceptionally clear coverage of mechanisms for catalysis, forces in aqueous solution, carbonyl- and acyl-group reactions, practical kinetics, more. 864pp. 5⅜ x 8½. 65460-5 Pa. $19.95

THE HISTORICAL BACKGROUND OF CHEMISTRY, Henry M. Leicester. Evolution of ideas, not individual biography. Concentrates on formulation of a coherent set of chemical laws. 260pp. 5⅜ x 8½. 61053-5 Pa. $8.95

A SHORT HISTORY OF CHEMISTRY (3rd edition), J. R. Partington. Classic exposition explores origins of chemistry, alchemy, early medical chemistry, nature of atmosphere, theory of valency, laws and structure of atomic theory, much more. 428pp. 5⅜ x 8½. (Available in U.S. only.) 65977-1 Pa. $12.95

Chemistry

GENERAL CHEMISTRY, Linus Pauling. Revised 3rd edition of classic first-year text by Nobel laureate. Atomic and molecular structure, quantum mechanics, statistical mechanics, thermodynamics correlated with descriptive chemistry. Problems. 992pp. 5⅜ x 8½. 65622-5 Pa. $19.95

Engineering

DE RE METALLICA, Georgius Agricola. The famous Hoover translation of greatest treatise on technological chemistry, engineering, geology, mining of early modern times (1556). All 289 original woodcuts. 638pp. 6¾ x 11. 60006-8 Pa. $21.95

FUNDAMENTALS OF ASTRODYNAMICS, Roger Bate et al. Modern approach developed by U.S. Air Force Academy. Designed as a first course. Problems, exercises. Numerous illustrations. 455pp. 5⅜ x 8½. 60061-0 Pa. $12.95

DYNAMICS OF FLUIDS IN POROUS MEDIA, Jacob Bear. For advanced students of ground water hydrology, soil mechanics and physics, drainage and irrigation engineering and more. 335 illustrations. Exercises, with answers. 784pp. 6⅛ x 9¼. 65675-6 Pa. $24.95

ANALYTICAL MECHANICS OF GEARS, Earle Buckingham. Indispensable reference for modern gear manufacture covers conjugate gear-tooth action, gear-tooth profiles of various gears, many other topics. 263 figures. 102 tables. 546pp. 5⅜ x 8½. 65712-4 Pa. $16.95

ADVANCED STRENGTH OF MATERIALS, J. P. Den Hartog. Superbly written advanced text covers torsion, rotating disks, membrane stresses in shells, much more. Many problems and answers. 388pp. 5⅜ x 8½. 65407-9 Pa. $11.95

MECHANICS, J. P. Den Hartog. A classic introductory text or refresher. Hundreds of applications and design problems illuminate fundamentals of trusses, loaded beams and cables, etc. 334 answered problems. 462pp. 5⅜ x 8½. 60754-2 Pa. $13.95

MECHANICAL VIBRATIONS, J. P. Den Hartog. Classic textbook offers lucid explanations and illustrative models, applying theories of vibrations to a variety of practical industrial engineering problems. Numerous figures. 233 problems, solutions. Appendix. Index. Preface. 436pp. 5⅜ x 8½. 64785-4 Pa. $13.95

STRENGTH OF MATERIALS, J. P. Den Hartog. Full, clear treatment of basic material (tension, torsion, bending, etc.) plus advanced material on engineering methods, applications. 350 answered problems. 323pp. 5⅜ x 8½. 60755-0 Pa. $10.95

A HISTORY OF MECHANICS, René Dugas. Monumental study of mechanical principles from antiquity to quantum mechanics. Contributions of ancient Greeks, Galileo, Leonardo, Kepler, Lagrange, many others. 671pp. 5⅜ x 8½. 65632-2 Pa. $18.95

STATISTICAL MECHANICS: Principles and Applications, Terrell L. Hill. Standard text covers fundamentals of statistical mechanics, applications to fluctuation theory, imperfect gases, distribution functions, more. 448pp. 5⅜ x 8½. 65390-0 Pa. $14.95

Engineering

THE VARIATIONAL PRINCIPLES OF MECHANICS, Cornelius Lanczos. Classic treatise provides graduate-level coverage of calculus of variations, equations of motion, relativistic mechanics, more. 418pp. 5⅜ x 8½. 65067-7 Pa. $14.95

THE VARIOUS AND INGENIOUS MACHINES OF AGOSTINO RAMELLI: A Classic Sixteenth-Century Illustrated Treatise on Technology, Agostino Ramelli. One of the most widely known and copied works on machinery in the 16th century. 194 detailed plates of water pumps, grain mills, cranes, more. 608pp. 9 x 12. 28180-9 Pa. $24.95

ORDINARY DIFFERENTIAL EQUATIONS AND STABILITY THEORY: An Introduction, David A. Sánchez. Brief, modern treatment. Linear equation, stability theory for autonomous and nonautonomous systems, etc. 164pp. 5⅜ x 8¼. 63828-6 Pa. $6.95

ROTARY-WING AERODYNAMICS, W. Z. Stepniewski. Clear, concise text covers aerodynamic phenomena of the rotor and offers guidelines for helicopter performance evaluation. Originally prepared for NASA. 537 figures. 640pp. 6⅛ x 9¼. 64647-5 Pa. $16.95

INTRODUCTION TO SPACE DYNAMICS, William Tyrrell Thomson. Comprehensive, classic introduction to space-flight engineering for advanced undergraduate and graduate students. Includes vector algebra, kinematics, transformation of coordinates. Bibliography. Index. 352pp. 5⅜ x 8½. 65113-4 Pa. $10.95

HISTORY OF STRENGTH OF MATERIALS, Stephen P. Timoshenko. Excellent survey with many references to the theories of elasticity and structure. 245 figures. 452pp. 5⅜ x 8½. 61187-6 Pa. $14.95

CONSTRUCTIONS AND COMBINATORIAL PROBLEMS IN DESIGN OF EXPERIMENTS, Damaraju Raghavarao. In-depth reference work examines orthogonal Latin squares, incomplete block designs, tactical configuration, partial geometry, much more. Abundant explanations, examples. 416pp. 5⅜ x 8¼. 65685-3 Pa. $10.95

Mathematics

HANDBOOK OF MATHEMATICAL FUNCTIONS WITH FORMULAS, GRAPHS, AND MATHEMATICAL TABLES, edited by Milton Abramowitz and Irene A. Stegun. Vast compendium: 29 sets of tables, some to as high as 20 places. 1,046pp. 8 x 10½. 61272-4 Pa. $29.95

CALCULUS REFRESHER FOR TECHNICAL PEOPLE, A. Albert Klaf. Covers important aspects of integral and differential calculus via 756 questions. 566 problems, most answered. 431pp. 5⅜ x 8½. 20370-0 Pa. $9.95

ASYMPTOTIC EXPANSIONS OF INTEGRALS, Norman Bleistein & Richard A. Handelsman. Best introduction to important field with applications in a variety of scientific disciplines. New preface. Problems. Diagrams. Tables. Bibliography. Index. 448pp. 5⅜ x 8½. 65082-0 Pa. $13.95

Mathematics

FAMOUS PROBLEMS OF GEOMETRY AND HOW TO SOLVE THEM, Benjamin Bold. Squaring the circle, trisecting the angle, duplicating the cube: learn their history, why they are impossible to solve, then solve them yourself. 128pp. 5⅜ x 8½. 24297-8 Pa. $6.95

VECTOR AND TENSOR ANALYSIS WITH APPLICATIONS, A. I. Borisenko and I. E. Tarapov. Concise introduction. Worked-out problems, solutions, exercises. 257pp. 5⅝ x 8¼. 63833-2 Pa. $10.95

THE ABSOLUTE DIFFERENTIAL CALCULUS (CALCULUS OF TENSORS), Tullio Levi-Civita. Great 20th-century mathematician's classic work on material necessary for grasp of theory of relativity. 452pp. 5⅜ x 8½. 63401-9 Pa. $14.95

AN INTRODUCTION TO ORDINARY DIFFERENTIAL EQUATIONS, Earl A. Coddington. A thorough and systematic first course in elementary differential equations for undergraduates in mathematics and science, with many exercises and problems (with answers). Index. 304pp. 5⅜ x 8½. 65942-9 Pa. $9.95

FOURIER SERIES AND ORTHOGONAL FUNCTIONS, Harry F. Davis. An incisive text combining theory and practical example to introduce Fourier series, orthogonal functions, and applications of the Fourier method to boundary-value problems. 570 exercises. Answers and notes. 416pp. 5⅜ x 8½. 65973-9 Pa. $13.95

COMPUTABILITY AND UNSOLVABILITY, Martin Davis. Classic graduate-level introduction to theory of computability, usually referred to as theory of recurrent functions. New preface and appendix. 288pp. 5⅜ x 8½. 61471-9 Pa. $8.95

ASYMPTOTIC METHODS IN ANALYSIS, N. G. de Bruijn. An inexpensive, comprehensive guide to asymptotic methods–the pioneering work that teaches by explaining worked examples in detail. Index. 224pp. 5⅜ x 8½. 64221-6 Pa. $9.95

ESSAYS ON THE THEORY OF NUMBERS, Richard Dedekind. Two classic essays by great German mathematician: on irrational numbers; and on transfinite numbers and properties of natural numbers. 115pp. 5⅜ x 8½. 21010-3 Pa. $6.95

APPLIED COMPLEX VARIABLES, John W. Dettman. Step-by-step coverage of fundamentals of analytic function theory–plus lucid exposition of five important applications: Potential Theory; Ordinary Differential Equations; Fourier Transforms; Laplace Transforms; Asymptotic Expansions. 66 figures. Exercises at chapter ends. 512pp. 5⅜ x 8½. 64670-X Pa. $14.95

INTRODUCTION TO LINEAR ALGEBRA AND DIFFERENTIAL EQUATIONS, John W. Dettman. Excellent text covers complex numbers, determinants, orthonormal bases, Laplace transforms, much more. Exercises with solutions. Undergraduate level. 416pp. 5⅜ x 8½. 65191-6 Pa. $12.95

MATHEMATICAL METHODS IN PHYSICS AND ENGINEERING, John W. Dettman. Algebraically based approach to vectors, mapping, diffraction, other topics in applied math. Also generalized functions, analytic function theory, more. Exercises. 448pp. 5⅜ x 8¼. 65649-7 Pa. $12.95

Mathematics

COMPLEX VARIABLES, Francis J. Flanigan. Unusual approach, delaying complex algebra till harmonic functions have been analyzed from real variable viewpoint. Includes problems with answers. 364pp. 5⅜ x 8½. 61388-7 Pa. $10.95

AN INTRODUCTION TO THE CALCULUS OF VARIATIONS, Charles Fox. Graduate-level text covers variations of an integral, isoperimetrical problems, least action, special relativity, approximations, more. References. 279pp. 5⅜ x 8½. 65499-0 Pa. $10.95

CATASTROPHE THEORY FOR SCIENTISTS AND ENGINEERS, Robert Gilmore. Advanced treatment describes mathematics of theory grounded in the work of Poincaré, R. Thom, other mathematicians. Also important applications to problems in mathematics, physics, chemistry, and engineering. 1981 edition. xvii + 666pp. 6⅛ x 9¼. 67539-4 Pa. $17.95

INTRODUCTION TO DIFFERENCE EQUATIONS, Samuel Goldberg. Exceptionally clear exposition of important discipline with applications to sociology, psychology, economics. Many illustrative examples; over 250 problems. 260pp. 5⅜ x 8½. 65084-7 Pa. $10.95

UNBOUNDED LINEAR OPERATORS: Theory and Applications, Seymour Goldberg. Classic presents systematic treatment of the theory of unbounded linear operators in normed linear spaces with applications to differential equations. Bibliography. 199pp. 5⅜ x 8½. 64830-3 Pa. $7.95

NUMERICAL METHODS FOR SCIENTISTS AND ENGINEERS, Richard Hamming. Classic text stresses frequency approach in coverage of algorithms, polynomial approximation, Fourier approximation, exponential approximation, other topics. Revised and enlarged 2nd edition. 721pp. 5⅜ x 8½. 65241-6 Pa. $17.95

POPULAR LECTURES ON MATHEMATICAL LOGIC, Hao Wang. Noted logician's lucid treatment of historical developments, set theory, model theory, recursion theory and constructivism, proof theory, more. 3 appendixes. Bibliography. 1981 edition. ix + 283pp. 5⅜ x 8½. 67632-3 Pa. $10.95

INTRODUCTION TO NUMERICAL ANALYSIS (2nd Edition), F. B. Hildebrand. Classic, fundamental treatment covers computation, approximation, interpolation, numerical differentiation and integration, other topics. 150 new problems. 669pp. 5⅜ x 8½. 65363-3 Pa. $16.95

THE FUNCTIONS OF MATHEMATICAL PHYSICS, Harry Hochstadt. Comprehensive treatment of orthogonal polynomials, hypergeometric functions, Hill's equation, much more. Bibliography. Index. 322pp. 5⅜ x 8½. 65214-9 Pa. $12.95

THREE PEARLS OF NUMBER THEORY, A. Y. Khinchin. Three compelling puzzles require proof of a basic law governing the world of numbers. Challenges concern van der Waerden's theorem, the Landau-Schnirelmann hypothesis and Mann's theorem, and a solution to Waring's problem. Solutions included. 64pp. 5⅜ x 8½. 40026-3 Pa. $4.95

Mathematics

THE PHILOSOPHY OF MATHEMATICS: An Introductory Essay, Stephan Körner. Surveys the views of Plato, Aristotle, Leibniz, and Kant concerning propositions and theories of applied and pure mathematics. Introduction. Two appendices. Index. 198pp. 5⅜ x 8½. 25048-2 Pa. $8.95

INTRODUCTORY REAL ANALYSIS, A.N. Kolmogorov and S. V. Fomin. Self-contained, evenly paced introduction to real and functional analysis. Some 350 problems. 403pp. 5⅜ x 8½. 61226-0 Pa. $12.95

APPLIED ANALYSIS, Cornelius Lanczos. Classic work on analysis and design of finite processes for approximating solution of analytical problems. Algebraic equations, matrices, harmonic analysis, quadrature methods, much more. 559pp. 5⅜ x 8½. 65656-X Pa. $16.95

AN INTRODUCTION TO ALGEBRAIC STRUCTURES, Joseph Landin. Superb self-contained text covers "abstract algebra": sets and numbers, theory of groups, theory of rings, much more. Numerous well-chosen examples, exercises. 247pp. 5⅜ x 8½. 65940-2 Pa. $10.95

SPECIAL FUNCTIONS, N. N. Lebedev. Famous Russian work treating more important special functions, with applications to specific problems of physics and engineering. 38 figures. 308pp. 5⅜ x 8½. 60624-4 Pa. $12.95

QUALITATIVE THEORY OF DIFFERENTIAL EQUATIONS, V. V. Nemytskii and V. V. Stepanov. Classic graduate-level text by two prominent Soviet mathematicians covers classical differential equations as well as topological dynamics and ergodic theory. Bibliographies. 523pp. 5⅜ x 8½. 65954-2 Pa. $14.95

NUMBER THEORY AND ITS HISTORY, Oystein Ore. Unusually clear, accessible introduction covers counting, properties of numbers, prime numbers, much more. Bibliography. 380pp. 5⅜ x 8½. 65620-9 Pa. $10.95

THEORY OF MATRICES, Sam Perlis. Outstanding text covering rank, nonsingularity, and inverses in connection with the development of canonical matrices under the relation of equivalence, and without the intervention of determinants. Includes exercises. 237pp. 5⅜ x 8½. 66810-X Pa. $8.95

INTRODUCTION TO ANALYSIS, Maxwell Rosenlicht. Unusually clear, accessible coverage of set theory, real number system, metric spaces, continuous functions, Riemann integration, multiple integrals, more. Wide range of problems. Undergraduate level. Bibliography. 254pp. 5⅜ x 8½. 65038-3 Pa. $9.95

MODERN NONLINEAR EQUATIONS, Thomas L. Saaty. Emphasizes practical solution of problems; covers seven types of equations. ". . . a welcome contribution to the existing literature...."–*Math Reviews*. 490pp. 5⅜ x 8½. 64232-1 Pa. $13.95

MATRICES AND LINEAR ALGEBRA, Hans Schneider and George Phillip Barker. Basic textbook covers theory of matrices and its applications to systems of linear equations and related topics such as determinants, eigenvalues and differential equations. Numerous exercises. 432pp. 5⅜ x 8½. 66014-1 Pa. $12.95

Mathematics

MATHEMATICS APPLIED TO CONTINUUM MECHANICS, Lee A. Segel. Analyzes models of fluid flow and solid deformation. For upper-level math, science and engineering students. 608pp. 5⅜ x 8½. 65369-2 Pa. $18.95

ELEMENTS OF REAL ANALYSIS, David A. Sprecher. Classic text covers fundamental concepts, real number system, point sets, functions of a real variable, Fourier series, much more. Over 500 exercises. 352pp. 5⅜ x 8½. 65385-4 Pa. $11.95

AN INTRODUCTION TO MATRICES, SETS AND GROUPS FOR SCIENCE STUDENTS, G. Stephenson. Concise, readable text introduces sets, groups, and most importantly, matrices to undergraduate students of physics, chemistry, and engineering. Problems. 164pp. 5⅜ x 8½. 65077-4 Pa. $7.95

SET THEORY AND LOGIC, Robert R. Stoll. Lucid introduction to unified theory of mathematical concepts. Set theory and logic seen as tools for conceptual understanding of real number system. 496pp. 5⅜ x 8¼. 63829-4 Pa. $14.95

TENSOR CALCULUS, J. L. Synge and A. Schild. Widely used introductory text covers spaces and tensors, basic operations in Riemannian space, non-Riemannian spaces, etc. 324pp. 5⅜ x 8¼. 63612-7 Pa. $11.95

ORDINARY DIFFERENTIAL EQUATIONS, Morris Tenenbaum and Harry Pollard. Exhaustive survey of ordinary differential equations for undergraduates in mathematics, engineering, science. Thorough analysis of theorems. Diagrams. Bibliography. Index. 818pp. 5⅜ x 8½. 64940-7 Pa. $19.95

CHALLENGING MATHEMATICAL PROBLEMS WITH ELEMENTARY SOLUTIONS, A. M. Yaglom and I. M. Yaglom. Over 170 challenging problems on probability theory, combinatorial analysis, points and lines, topology, convex polygons, many other topics. Solutions. Total of 445pp. 5⅜ x 8½. Two-vol. set.
Vol. I: 65536-9 Pa. $8.95
Vol. II: 65537-7 Pa. $8.95

INTEGRAL EQUATIONS, F. G. Tricomi. Authoritative, well-written treatment of extremely useful mathematical tool with wide applications. Volterra Equations, Fredholm Equations, much more. Advanced undergraduate to graduate level. Exercises. Bibliography. 238pp. 5⅜ x 8½. 64828-1 Pa. $8.95

FOURIER SERIES, Georgi P. Tolstov. Translated by Richard A. Silverman. A valuable addition to the literature on the subject, moving clearly from subject to subject and theorem to theorem. 107 problems, answers. 336pp. 5⅜ x 8½. 63317-9 Pa. $11.95

DISTRIBUTION THEORY AND TRANSFORM ANALYSIS: An Introduction to Generalized Functions, with Applications, A. H. Zemanian. Provides basics of distribution theory, describes generalized Fourier and Laplace transformations. Numerous problems. 384pp. 5⅜ x 8½. 65479-6 Pa. $13.95

CALCULUS OF VARIATIONS, Robert Weinstock. Basic introduction covering isoperimetric problems, theory of elasticity, quantum mechanics, electrostatics, etc. Exercises throughout. 326pp. 5⅜ x 8½. 63069-2 Pa. $9.95

Mathematics

THE CONTINUUM: A Critical Examination of the Foundation of Analysis, Hermann Weyl. Classic of 20th-century foundational research deals with the conceptual problem posed by the continuum. 156pp. 5⅜ x 8½. 67982-9 Pa. $8.95

A SURVEY OF NUMERICAL MATHEMATICS, David M. Young and Robert Todd Gregory. Broad self-contained coverage of computer-oriented numerical algorithms for solving various types of mathematical problems in linear algebra, ordinary and partial, differential equations, much more. Exercises. Total of 1,248pp. 5⅜ x 8½. Two volumes. Vol. I: 65691-8 Pa. $16.95
Vol. II: 65692-6 Pa. $16.95

INTRODUCTION TO PARTIAL DIFFERENTIAL EQUATIONS WITH APPLICATIONS, E. C. Zachmanoglou and Dale W. Thoe. Essentials of partial differential equations applied to common problems in engineering and the physical sciences. Problems and answers. 416pp. 5⅜ x 8½. 65251-3 Pa. $11.95

THE THEORY OF GROUPS, Hans J. Zassenhaus. Well-written graduate-level text acquaints reader with group-theoretic methods and demonstrates their usefulness in mathematics. Axioms, the calculus of complexes, homomorphic mapping, *p*-group theory, more. Excellent proofs. 276pp.5⅜ x 8½. 40922-8 Pa. $12.95

GENERALIZED INTEGRAL TRANSFORMATIONS, A.H. Zemanian. Graduate-level study of recent generalizations of the Laplace, Mellin, Hankel, K. Weierstrass, convolution and other simple transformations. Bibliography. 320pp. 5⅜ x 8½. 65375-7 Pa. $8.95

Math–Decision Theory, Statistics, Probability

PROBABILITY: An Introduction, Samuel Goldberg. Excellent basic text covers set theory, probability theory for finite sample spaces, binomial theorem, much more. 360 problems. Bibliographies. 322pp. 5⅜ x 8½. 65252-1 Pa. $11.95

ELEMENTARY DECISION THEORY, Herman Chernoff and Lincoln E. Moses. Clear introduction to statistics and statistical theory covers data processing, probability and random variables, testing hypotheses, much more. Exercises. 364pp. 5⅜ x 8½. 65218-1 Pa. $12.95

STATISTICS MANUAL, Edwin L. Crow et al. Comprehensive, practical collection of classical and modern methods prepared by U.S. Naval Ordnance Test Station. Stress on use. Basics of statistics assumed. 288pp. 5⅜ x 8½. 60599-X Pa. $8.95

SOME THEORY OF SAMPLING, William Edwards Deming. Analysis of the problems, theory, and design of sampling techniques for social scientists, industrial managers, and others. xvii + 602pp. 5⅜ x 8½. 64684-X Pa. $16.95

STATISTICAL ADJUSTMENT OF DATA, W. Edwards Deming. Introduction to basic concepts of statistics, curve fitting, least squares solution, conditions without parameter, conditions containing parameters. 26 exercises worked out. 271pp. 5⅜ x 8½. 64685-8 Pa. $9.95

Math–Decision Theory, Statistics, Probability

LINEAR PROGRAMMING AND ECONOMIC ANALYSIS, Robert Dorfman, Paul A. Samuelson, and Robert M. Solow. First comprehensive treatment of linear programming in standard economic analysis. Game theory, modern welfare economics, Leontief input-output, more. 525pp. 5⅜ x 8½. 65491-5 Pa. $17.95

DICTIONARY/OUTLINE OF BASIC STATISTICS, John E. Freund and Frank J. Williams. A clear concise dictionary of over 1,000 statistical terms and an outline of statistical formulas covering probability, nonparametric tests, much more. 208pp. 5⅜ x 8½. 66796-0 Pa. $8.95

GAMES AND DECISIONS: Introduction and Critical Survey, R. Duncan Luce, and Howard Raiffa. Superb nontechnical introduction to game theory, primarily applied to social sciences. Utility theory, zero-sum games, n-person games, decision-making, much more. Bibliography. 509pp. 5⅜ x 8½. 65943-7 Pa. $14.95

FIFTY CHALLENGING PROBLEMS IN PROBABILITY WITH SOLUTIONS, Frederick Mosteller. Remarkable puzzlers, graded in difficulty, illustrate elementary and advanced aspects of probability. Detailed solutions. 88pp. 5⅜ x 8½. 65355-2 Pa. $5.95

OPTIMIZATION THEORY WITH APPLICATIONS, Donald A. Pierre. Broad spectrum approach to important topic. Classical theory of minima and maxima, calculus of variations, simplex technique and linear programming, more. Many problems, examples. 640pp. 5⅜ x 8½. 65205-X Pa. $17.95

PROBABILITY THEORY: A Concise Course, Y. A. Rozanov. Highly readable, self-contained introduction covers combination of events, dependent events, Bernoulli trials, etc. 148pp. 5⅜ x 8¼. 63544-9 Pa. $8.95

STATISTICAL METHOD FROM THE VIEWPOINT OF QUALITY CONTROL, Walter A. Shewhart. Important text explains regulation of variables, uses of statistical control to achieve quality control in industry, agriculture, other areas. 192pp. 5⅜ x 8½. 65232-7 Pa. $8.95

THE COMPLEAT STRATEGYST: Being a Primer on the Theory of Games of Strategy, J. D. Williams. Highly entertaining classic describes, with many examples, how to select best strategies in conflict situations. 268pp. 5⅜ x 8½. 25101-2 Pa. $9.95

Math–Geometry and Topology

ELEMENTARY CONCEPTS OF TOPOLOGY, Paul Alexandroff. Elegant, intuitive approach to topology from set-theoretic topology to Betti groups; how concepts of topology are useful in math and physics. 57pp. 5⅜ x 8½. 60747-X Pa. $4.95

COMBINATORIAL TOPOLOGY, P. S. Alexandrov. Clearly written, well-organized, three-part text begins by dealing with certain classic problems without using the formal techniques of homology theory and advances to the central concept, the Betti groups. Numerous detailed examples. 654pp. 5⅜ x 8½. 40179-0 Pa. $18.95

Math–Geometry and Topology

EXPERIMENTS IN TOPOLOGY, Stephen Barr. Classic, lively explanation of one of the byways of mathematics. Klein bottles, Moebius strips, projective planes, map coloring, problem of the Koenigsberg bridges, much more, described with clarity and wit. 43 figures. 210pp. 5⅜ x 8½. 25933-1 Pa. $8.95

THE GEOMETRY OF RENÉ DESCARTES, René Descartes. The great work founded analytical geometry. Original French text, Descartes's own diagrams, together with definitive Smith-Latham translation. 244pp. 5⅜ x 8½. 60068-8 Pa. $9.95

THE THIRTEEN BOOKS OF EUCLID'S ELEMENTS, translated with introduction and commentary by Sir Thomas L. Heath. Definitive edition. Textual and linguistic notes, mathematical analysis. 2,500 years of critical commentary. Unabridged. 1,414pp. 5⅜ x 8½. Three-vol. set. Vol. I: 60088-2 Pa. $11.95
Vol. II: 60089-0 Pa. $10.95
Vol. III: 60090-4 Pa. $12.95

CONFORMAL MAPPING ON RIEMANN SURFACES, Harvey Cohn. Lucid, insightful book presents ideal coverage of subject. 334 exercises make book perfect for self-study. 55 figures. 352pp. 5⅝ x 8¼. 64025-6 Pa. $11.95

DIFFERENTIAL GEOMETRY, Heinrich W. Guggenheimer. Local differential geometry as an application of advanced calculus and linear algebra. Curvature, transformation groups, surfaces, more. Exercises. 62 figures. 378pp. 5⅜ x 8½.
63433-7 Pa. $11.95

CURVATURE AND HOMOLOGY: Enlarged Edition, Samuel I. Goldberg. Revised edition examines topology of differentiable manifolds; curvature, homology of Riemannian manifolds; compact Lie groups; complex manifolds; curvature, homology of Kaehler manifolds. New Preface. Four new appendixes. 416pp. 5⅜ x 8½.
40207-X Pa. $14.95

TOPOLOGY, John G. Hocking and Gail S. Young. Superb one-year course in classical topology. Topological spaces and functions, point-set topology, much more. Examples and problems. Bibliography. Index. 384pp. 5⅝ x 8¼. 65676-4 Pa. $11.95

THE FOUR-COLOR PROBLEM: Assaults and Conquest, Thomas L. Saaty and Paul G. Kainen. Engrossing, comprehensive account of the century-old combinatorial topological problem, its history and solution. Bibliographies. Index. 110 figures. 228pp. 5⅜ x 8½. 65092-8 Pa. $7.95

GEOMETRY OF COMPLEX NUMBERS, Hans Schwerdtfeger. Illuminating, widely praised book on analytic geometry of circles, the Moebius transformation, and two-dimensional non-Euclidean geometries. 200pp. 5⅝ x 8¼. 63830-8 Pa. $8.95

LECTURES ON CLASSICAL DIFFERENTIAL GEOMETRY, Second Edition, Dirk J. Struik. Excellent brief introduction covers curves, theory of surfaces, fundamental equations, geometry on a surface, conformal mapping, other topics. Problems. 240pp. 5⅜ x 8½. 65609-8 Pa. $9.95

Math–History of

A SHORT ACCOUNT OF THE HISTORY OF MATHEMATICS, W. W. Rouse Ball. One of clearest, most authoritative surveys from the Egyptians and Phoenicians through 19th-century figures such as Grassman, Galois, Riemann. Fourth edition. 522pp. 5⅜ x 8½. 20630-0 Pa. $13.95

THE HISTORY OF THE CALCULUS AND ITS CONCEPTUAL DEVELOPMENT, Carl B. Boyer. Origins in antiquity, medieval contributions, work of Newton, Leibniz, rigorous formulation. Treatment is verbal. 346pp. 5⅜ x 8½. 60509-4 Pa. $9.95

THE HISTORICAL ROOTS OF ELEMENTARY MATHEMATICS, Lucas N. H. Bunt, Phillip S. Jones, and Jack D. Bedient. Fundamental underpinnings of modern arithmetic, algebra, geometry and number systems derived from ancient civilizations. 320pp. 5⅜ x 8½. 25563-8 Pa. $9.95

GAMES, GODS & GAMBLING: A History of Probability and Statistical Ideas, F. N. David. Episodes from the lives of Galileo, Fermat, Pascal, and others illustrate this fascinating account of the roots of mathematics. Features thought-provoking references to classics, archaeology, biography, poetry. 1962 edition. 304pp. 5⅜ x 8½. (Available in U.S. only.) 40023-9 Pa. $9.95

HISTORY OF MATHEMATICS, David E. Smith. Nontechnical survey from ancient Greece and Orient to late 19th century; evolution of arithmetic, geometry, trigonometry, calculating devices, algebra, the calculus. 362 illustrations. 1,355pp. 5⅜ x 8½. Two-vol. set. Vol. I: 20429-4 Pa. $13.95
Vol. II: 20430-8 Pa. $14.95

A CONCISE HISTORY OF MATHEMATICS, Dirk J. Struik. The best brief history of mathematics. Stresses origins and covers every major figure from ancient Near East to 19th century. 41 illustrations. 195pp. 5⅜ x 8½. 60255-9 Pa. $8.95

Meteorology

PRINCIPLES OF METEOROLOGICAL ANALYSIS, Walter J. Saucier. Highly respected, abundantly illustrated classic reviews atmospheric variables, hydrostatics, static stability, various analyses (scalar, cross-section, isobaric, isentropic, more). For intermediate meteorology students. 454pp. 6⅛ x 9¼. 65979-8 Pa. $14.95

LIGHTNING, Martin A. Uman. Revised, updated edition of classic work on the physics of lightning. Phenomena, terminology, measurement, photography, spectroscopy, thunder, more. Reviews recent research. Bibliography. Indices. 320pp. 5⅜ x 8¼. 64575-4 Pa. $10.95

Physics

OPTICAL RESONANCE AND TWO-LEVEL ATOMS, L. Allen and J. H. Eberly. Clear, comprehensive introduction to basic principles behind all quantum optical resonance phenomena. 53 illustrations. Preface. Index. 256pp. 5⅜ x 8½.
65533-4 Pa. $10.95

Physics

ULTRASONIC ABSORPTION: An Introduction to the Theory of Sound Absorption and Dispersion in Gases, Liquids and Solids, A. B. Bhatia. Standard reference in the field provides a clear, systematically organized introductory review of fundamental concepts for advanced graduate students, research workers. Numerous diagrams. Bibliography. 440pp. 5⅜ x 8½. 64917-2 Pa. $11.95

QUANTUM THEORY, David Bohm. This advanced undergraduate-level text presents the quantum theory in terms of qualitative and imaginative concepts, followed by specific applications worked out in mathematical detail. Preface. Index. 655pp. 5⅜ x 8½. 65969-0 Pa. $15.95

ATOMIC PHYSICS (8th edition), Max Born. Nobel laureate's lucid treatment of kinetic theory of gases, elementary particles, nuclear atom, wave-corpuscles, atomic structure and spectral lines, much more. Over 40 appendices, bibliography. 495pp. 5⅜ x 8½. 65984-4 Pa. $14.95

AN INTRODUCTION TO HAMILTONIAN OPTICS, H. A. Buchdahl. Detailed account of the Hamiltonian treatment of aberration theory in geometrical optics. Many classes of optical systems defined in terms of the symmetries they possess. Problems with detailed solutions. 1970 edition. xv + 360pp. 5⅜ x 8½. 67597-1 Pa. $10.95

THIRTY YEARS THAT SHOOK PHYSICS: The Story of Quantum Theory, George Gamow. Lucid, accessible introduction to influential theory of energy and matter. Careful explanations of Dirac's antiparticles, Bohr's model of the atom, much more. 12 plates. Numerous drawings. 240pp. 5⅜ x 8½. 24895-X Pa. $8.95

ELECTRONIC STRUCTURE AND THE PROPERTIES OF SOLIDS: The Physics of the Chemical Bond, Walter A. Harrison. Innovative text offers basic understanding of the electronic structure of covalent and ionic solids, simple metals, transition metals, and their compounds. Problems. 1980 edition. 582pp. 6⅛ x 9¼. 66021-4 Pa. $19.95

HYDRODYNAMIC AND HYDROMAGNETIC STABILITY, S. Chandrasekhar. Lucid examination of the Rayleigh-Benard problem; clear coverage of the theory of instabilities causing convection. 704pp. 5⅜ x 8¼. 64071-X Pa. $17.95

INVESTIGATIONS ON THE THEORY OF THE BROWNIAN MOVEMENT, Albert Einstein. Five papers (1905–8) investigating dynamics of Brownian motion and evolving elementary theory. Notes by R. Fürth. 122pp. 5⅜ x 8½. 60304-0 Pa. $5.95

THE PHYSICS OF WAVES, William C. Elmore and Mark A. Heald. Fine overview of classical wave theory. Acoustics, optics, electromagnetic radiation, more. Ideal as classroom text or for self-study. Problems. 477pp. 5⅜ x 8½. 64926-1 Pa. $14.95

PHYSICAL PRINCIPLES OF THE QUANTUM THEORY, Werner Heisenberg. Nobel Laureate discusses quantum theory, uncertainty, wave mechanics, work of Dirac, Schroedinger, Compton, Wilson, Einstein, etc. 184pp. 5⅜ x 8½. 60113-7 Pa. $8.95

Physics

ATOMIC SPECTRA AND ATOMIC STRUCTURE, Gerhard Herzberg. One of best introductions; especially for specialist in other fields. Treatment is physical rather than mathematical. 80 illustrations. 257pp. 5⅜ x 8½. 60115-3 Pa. $7.95

AN INTRODUCTION TO STATISTICAL THERMODYNAMICS, Terrell L. Hill. Excellent basic text offers wide-ranging coverage of quantum statistical mechanics, systems of interacting molecules, quantum statistics, more. 523pp. 5⅜ x 8½. 65242-4 Pa. $14.95

THEORETICAL PHYSICS, Georg Joos, with Ira M. Freeman. Classic overview covers essential math, mechanics, electromagnetic theory, thermodynamics, quantum mechanics, nuclear physics, other topics. First paperback edition. xxiii+885pp. 5⅜ x 8½. 65227-0 Pa. $24.95

BOUNDARY VALUE PROBLEMS OF HEAT CONDUCTION, M. Necati Özisik. Systematic, comprehensive treatment of modern mathematical methods of solving problems in heat conduction and diffusion. Numerous examples and problems. Selected references. Appendices. 505pp. 5⅜ x 8½. 65990-9 Pa. $12.95

PROBLEMS AND SOLUTIONS IN QUANTUM CHEMISTRY AND PHYSICS, Charles S. Johnson Jr. and Lee G. Pedersen. Unusually varied problems, detailed solutions in coverage of quantum mechanics, wave mechanics, angular momentum, molecular spectroscopy, scattering theory, more. 280 problems plus 139 supplementary exercises. 430pp. 6½ x 9¼. 65236-X Pa. $14.95

THEORETICAL SOLID STATE PHYSICS, Vol. 1: Perfect Lattices in Equilibrium; Vol. II: Non-Equilibrium and Disorder, William Jones and Norman H. March. Monumental reference work covers fundamental theory of equilibrium properties of perfect crystalline solids, nonequilibrium properties, defects, and disordered systems. Appendices. Problems. Preface. Diagrams. Index. Bibliography. Total of 1,301pp. 5⅜ x 8½. Two volumes. Vol. I: 65015-4 Pa. $16.95
Vol. II: 65016-2 Pa. $16.95

A TREATISE ON ELECTRICITY AND MAGNETISM, James Clerk Maxwell. Important foundation work of modern physics. Brings to final form Maxwell's theory of electromagnetism and rigorously derives his general equations of field theory. 1,084pp. 5⅜ x 8½. Two-vol. set. Vol. I: 60636-8 Pa. $14.95
Vol. II: 60637-6 Pa. $14.95

OPTICKS, Sir Isaac Newton. Newton's own experiments with spectroscopy, colors, lenses, reflection, refraction, etc., in language the layman can follow. Foreword by Albert Einstein. 532pp. 5⅜ x 8½. 60205-2 Pa. $13.95

THEORY OF ELECTROMAGNETIC WAVE PROPAGATION, Charles Herach Papas. Graduate-level study discusses the Maxwell field equations, radiation from wire antennas, the Doppler effect and more. xiii + 244pp. 5⅜ x 8½. 65678-0 Pa. $9.95

METHODS OF THERMODYNAMICS, Howard Reiss. Outstanding text focuses on physical technique of thermodynamics, problem areas, and significance and use of thermodynamic potential. 1965 edition. 238pp. 5⅜ x 8½. 69445-3 Pa. $8.95

Physics

INTRODUCTION TO QUANTUM MECHANICS With Applications to Chemistry, Linus Pauling and E. Bright Wilson Jr. Classic undergraduate text by Nobel Prize winner applies quantum mechanics to chemical and physical problems. Numerous tables and figures enhance the text. Chapter bibliographies. Appendices. Index. 468pp. 5⅜ x 8½. 64871-0 Pa. $13.95

TENSOR ANALYSIS FOR PHYSICISTS, J. A. Schouten. Concise exposition of the mathematical basis of tensor analysis, integrated with well-chosen physical examples of the theory. Exercises. Index. Bibliography. 289pp. 5⅜ x 8½. 65582-2 Pa. $13.95

RELATIVITY IN ILLUSTRATIONS, Jacob T. Schwartz. Clear nontechnical treatment makes relativity more accessible than ever before. Over 60 drawings illustrate concepts more clearly than text alone. Only high school geometry needed. Bibliography. 128pp. 6⅛ x 9¼. 25965-X Pa. $7.95

THE ELECTROMAGNETIC FIELD, Albert Shadowitz. Comprehensive undergraduate text covers basics of electric and magnetic fields, builds up to electromagnetic theory. Also related topics, including relativity. Over 900 problems. 768pp. 5⅝ x 8¼. 65660-8 Pa. $19.95

GREAT EXPERIMENTS IN PHYSICS: Firsthand Accounts from Galileo to Einstein, edited by Morris H. Shamos. 25 crucial discoveries: Newton's laws of motion, Chadwick's study of the neutron, Hertz on electromagnetic waves, more. Original accounts clearly annotated. 370pp. 5⅜ x 8½. 25346-5 Pa. $12.95

RELATIVITY, THERMODYNAMICS AND COSMOLOGY, Richard C. Tolman. Landmark study extends thermodynamics to special, general relativity; alsoapplications of relativistic mechanics, thermodynamics to cosmological models. 501pp. 5⅜ x 8½. 65383-8 Pa. $15.95

LIGHT SCATTERING BY SMALL PARTICLES, H. C. van de Hulst. Comprehensive treatment including full range of useful approximation methods for researchers in chemistry, meteorology and astronomy. 44 illustrations. 470pp. 5⅜ x 8½. 64228-3 Pa. $14.95

STATISTICAL PHYSICS, Gregory H. Wannier. Classic text combines thermodynamics, statistical mechanics and kinetic theory in one unified presentation of thermal physics. Problems with solutions. Bibliography. 532pp. 5⅜ x 8½. 65401-X Pa. $14.95

Prices subject to change without notice.

Available at your book dealer or write for free Dover Mathematics and Science Catalog (59065-8) to Dept. GI, Dover Publications, Inc., 31 East 2nd St., Mineola, NY 11501. Dover publishes more than 400 books each year on science, elementary and advanced mathematics, biology, music, art, literature, history, social sciences, and other subjects.